Social Spatialization in a Turkish Squatter Settlement

European University Studies

Europäische Hochschulschriften
Publications Universitaires Européennes

Series XXII
Sociology

Reihe XXII Série XXII
Soziologie
Sociologie

Vol./Bd. 427

PETER LANG
Frankfurt am Main · Berlin · Bern · Bruxelles · New York · Oxford · Wien

Neslihan Demirtaş

Social Spatialization in a Turkish Squatter Settlement

The Dualism of Strategy and Tactic Reconsidered

PETER LANG
Internationaler Verlag der Wissenschaften

Bibliographic Information published by the Deutsche Nationalbibliothek
The Deutsche Nationalbibliothek lists this publication in the Deutsche Nationalbibliografie; detailed bibliographic data is available in the internet at <http://www.d-nb.de>.

ISSN 0721-3379
ISBN 978-3-631-57887-2

www.peterlang.de

ACKNOWLEDGEMENTS

This book developed out of doctoral research at Bilkent University, Department of Political Science. The study would not have been possible without the financial assistance and encouragement of my department, "Turkish Academy of Sciences Fellowship Program for Integrated Doctoral Studies in Turkey and Abroad in the Social Sciences and Humanities", "FURS (Foundation for Urban and Regional Studies) Research Grant" at University of Essex and "ARIT (American Research Institute in Turkey) Research Grant".

My greatest debt is to Banu Helvacıoğlu for her exceptional intellectual guidance, warm and encouraging emotional support and for opening my eyes to many invaluable and vibrant ideas and questions during my entire study. I also would like to thank Ahmet İçduygu for his generous academic guidance and support from the very initial design of the research. Alev Çınar, Güliz Ger, Nedim Karakayalı and Simon Wigley contributed to this study with their constructive criticisms from which I benefited a lot.

I would also like to mention my special gratitude to E. Fuat Keyman, Sibel Kalaycıoğlu and Helga Rittersberger Tılıç who have always been there to offer unconditional and invaluable intellectual and emotional help to me. I am indebted to my friend Işık Gürleyen for her rigorous reading and criticism. Working at İzmir University of Economics during this book project has given me the courage to complete my study. For this opportunity and endorsement, I would like to thank Orhan Tekelioğlu. The discussions I have made with my dear friends H. Ertuğ Tombuş and Seher Şen and their insights contributed a lot to the theoretical framework of the book. I am indebted to Ian Manners and my dear colleagues Margaret Lynch and Trushna Parekh for making the days I spent at the Department of Geography of University of Texas at Austin intellectually productive.

I am deeply grateful for the meticulous efforts of my dear friend Derya Irkdaş who restlessly formatted the script of this book. A special thanks goes to the preliminary proofreadings of Nesta Parry and Simon Mumford who refined the text with their engagement.

Last but not least, I'd like to mention my special thanks to all the respondents of the research, inhabitants of Boğaziçi, for opening their hearts and showing unconditional hospitality. Selvi, Ali, Ramazan and Haydar deserve my deep indebted-ness for walking with me long hours, introducing me many migrants and accepting me as a member of their families.

To my beloved mum, dad, Manfred, Nurhan and Nihan.

TABLE OF CONTENTS

CHAPTER 1 - Introduction

1.1 Conceptual Framework

Gecekondu is one of enduring problems in Turkey having political, societal and economic repercussions. This chapter starts with a brief analysis of various representations of *gecekondu* settlements in public debate. In order to have a better understanding of these various representations, a brief explanation of socio-spatial transformations in *gecekondu* settlements over time will be provided. The overall purpose is to introduce conceptual tools in accordance with the basic research objectives that have guided this book project. The spatial conceptual and methodological framework that will be employed throughout the book will also be made explicit within the context of this chapter.

1.1.1 Introduction

There has been a wide debate since the initial emergence of *gecekondu* (squatter) settlements in the largest cities of Turkey, namely Ankara, İstanbul and İzmir in the 1940s. *Gecekondu* literary means "built-overnight" and *gecekondulu* refers to people living in *gecekondu* settlements. Initially, the concept of *gecekondu* referred to houses or settlements constructed on state or privately owned land without planning and / or construction permission, through the efforts of rural migrants to urban areas and their *hemşehris*.[1]

In time, the meaning, attributes and status of *gecekondu* have changed. The early "*gecekondu*" neighborhoods have become legalized over time. Therefore, the classical definition of *gecekondu* does not apply to these settlements anymore. Especially since the early 1990s, *gecekondu* is no longer a concept signifying all the settlements of low-income people in the periphery of the cities. Besides, the newly emerging illegal residential development in the 1990s, which occurred mostly as a consequence of forced migration[2], hardly reflects the socio-spatial qualities of classical "*gecekondu*" settlements.

---

1 *Hemşehri* refers to the people of the same hometown. The concept *hemşehri* in daily usage means people of the same village or the same province.

2 Forced migration is either the forced deportation of the villagers, being Kurd in ethnic origin by state security forces or their unprepared sudden decision of rural-to-urban migration as a consequence of the conflicts between state security forces and PKK, a separatist group seeking to establish an independent Kurdish State in the southeastern part of Turkey. Forced migration dominated rural-to-urban migration starting from the early 1990s.

Parallel to the socio-spatial transformations in low-income settlements, the general public approach that have been defined by social actors having different agendas with regard to representation and definition of these low-income settlements have also been transformed over time. The different representation of these settlements in the public perception by diverse actors has affected *gecekondu* policy, where the general strategical parameters of dealing with the *gecekondu* issue have been defined. Diverse legal means to deal with the unlawful aspect of *gecekondu* settlements, the problem definitions in relation to *gecekondu* growth in political, academic and journalistic circles, planning policy orientations and all the attempts to represent *gecekondu* space and population are considered within the context of the strategical acts in dealing with *gecekondu* space in this study. To exemplify the tactical acts of the migrants in dealing with these strategical acts and to show the intertwined nature of the tactical and strategical realms are quite critical in understanding the nature of urbanization in Turkey. The tactical acts of migrants include all the critical decisions, actions and forms of resistance in order to survive both in relation to the hardships of the local context and to the strategical acts and interventions in *gecekondu* space. Consequently, the book will be an endeavor which will present a local history of a *gecekondu* neighborhood considering the intertwined nature of the strategical and tactical realms with an insider perspective. However, before making the research questions explicit, it will be helpful to define the problems of the strategical realm in dealing with *gecekondu* spaces over time. Thus, the following operational questions will help to define the context of the strategical realm, which are:

- What kind of a public discourse[3] regarding low-income settlements has been constructed over time?
- Who have dominated the definition of public discourse in relation to these settlements?
- What kind of changes and continuities can be identified over time with regard to this public discourse?

Although these questions will be discussed at length in the following chapter, it is necessary to briefly address them in order to highlight the relevancy of the research objectives of this work. At the outset it is also important to note that *gecekondu* and *gecekondulu* become objects in the eyes of various societal actors, who internalize different subject positions defining the role and

3 Public discourse here does not only refer to the representations and narratives on *gecekondu* issue in public debate but also to the legal or political practices in dealing with and defining *gecekondu* space.

status of *gecekondu*. The subjective concerns related to *gecekondu* settlements show variations for different actors who define the public agenda.

The interest of politicians in the *gecekondu* development is twofold. First, they perceive *gecekondus* as a potential source of votes. Second, politicians have a responsibility to solve the problems in these settlements as part of urban problems. There have also been numerous academic works on *gecekondu* settlements in different fields. Sociological studies dominate these academic works but *gecekondu* has become a topic of political science, economy, and urban planning as well, and journalists have often brought the problems related to these settlements into public attention by using all the means of the mass media.

Although there are variations in these different actors' motives in dealing with *gecekondu* settlements, a modernist-elitist perception has dominated their approaches to *gecekondu* settlements and *gecekondulu* to a large extent. In other words, modernist-elitist perspective, which defines public perception about *gecekondu* settlements, refers mostly to the credents of urban middle classes. The groups that have constructed public discourse about *gecekondu* settlements and *gecekondulu* are referred to as "urbanites", at various points in this book in order to reflect the common position these groups have internalized vis-à-vis *gecekondulus*. The term, "urbanite", is borrowed from Meral Özbek[4], signifying a loose group consisting of bourgeoisie, state bureaucrats, the urban middle classes, Kemalist intellectuals, and some radical intellectuals who define the parameters of urbanity mainly with reference to *gecekondulu*.

Public discourse to a large extent has been defined by the impositions of "urbanites" and thus has been shaped by a strong elitist perspective. Therefore *gecekondu* neighborhoods and their "way of life" have been conventionally portrayed in Turkish public understanding as the main problem and an impediment against modernization. The public discourse has taken different forms and been acknowledged by different urban groups over time depending on the macro political, social and economic context of the country and the socio-spatial transformations of *gecekondu* space(s) in this context.

In this study, the course of *gecekondu* settlements is divided into three main periods in accordance with their changing socio-spatial context. The first period in *gecekondu* history overlaps with the years of transition to multiparty politics in Turkey. The initial encounters with *gecekondu* settlements in the periphery of cities and their spread occur as a result of the initial populist of experiences of multi-party politics. In the 1940s *gecekondu* settlements

---

4 Özbek, Meral (1997), "Arabesk Culture: A Case of Modernization and Popular Identity", in Bozdoğan, S. and Kasaba, R. (eds.), *Rethinking the Modern Project in Turkey*, Seattle: Washington University Press, p.228.

first entered the public agenda with the initiation of migration from rural to urban areas, most of all to Ankara and İstanbul. During those years, the public policy towards these settlements started to take form. The Republican Regime's strategy of creating a modernist urban space that took Western cities and lifestyle as its main reference point in the late 1920s and early 1930s determined the content of elitist perception. This elitist understanding had in turn defined public discourse with regard to *gecekondu* settlements in this period of early encounters with *gecekondu* settlements.

The act of imposing a Western way of life and existence on society supposes ideal templates of "urbanity". The main act of the Republican Regime in the service of this ideal was to reconstruct Ankara, capital city of Turkey as a planned and modernist city. The city was planned and constructed in a twofold fashion as the socio-spatial representation of that ideal. First of all, the physical space of the city was constructed and planned with a strong reference to Western European cities. Secondly, in congruence with this physical planning, the social space of the city was constructed and implemented by taking the "urban way of life" in Western European cities as its main model. The construction of modernist social space was realized by making "urban lifestyle" visible to public. The modern lifestyle was modeled and demonstrated by the newly arrived elites of the city, namely the families of state bureaucrats, intellectuals and representatives of Western countries. This process had defined the parameters of being an "urbanite" in the early years of the reconstruction of the city of Ankara.

In spite of all planning attempts under the rule of a Single Party regime, namely the Republican People's Party, *gecekondu* settlements in the periphery and slum-like residential areas inhabiting seasonal workers at the traditional centers of Ankara and İstanbul initially became visible in the early 1940s. Yet, these settlements were few in number and constructed primarily for the purpose of temporary shelter. The intellectuals, journalists, bureaucrats, and academics of the early Republican Regime who had been socialized via the Republican ideals shaped public discourse in those years and delineated the terms of "urbanity" by mainly contrasting it with its "other", namely *gecekondu* spaces. Furthermore, the ideal templates of urbanity and urban space have been continuously defined by taking examples from the West. The "other" of these ideal templates has been continuously defined via a depiction of spontaneous spaces (*gecekondu* neighborhoods) emerging within the system, inconsistent with the ideals of modernization. In this respect, public discourse usually neglects the embeddedness of socio-spatial transformations of these space(s) in the changing context of the country.

The modern life style, as imposed by the Republican Regime, from the first years, encountered serious inconsistencies with the context of everyday life in the large cities of the country. The public perception blames rural-to-urban migrants as the main reason behind the intensification of these irreconcilabilities. It attributes these irreconcilabilities to the intrinsic marginal qualities of *gecekondu* settlements rather than the impediments embedded in the attempt to impose a context-free modern space from the top down. A certain attitude of neglect and unease on the part of the Republican elites regarding these spaces emerged. Alongside this uneasiness there had also been a wishful and optimistic public perception that was based on a belief on the transitory qualities of these settlements. Since these were the initial years of migration and these settlements were few in number, they were considered as transitory structures that would be integrated into cities in time. In addition, the rural-to-urban migration was also considered as reversible by means of macro economic and political strategies.

Contrary to the early estimates, the rate of rural-to-urban migration increased with the populist concerns of political actors of multi-party period that was initiated by Democrat Party's (*DP*) coming to power in 1950. The politicians began to realize the increasing effect of *gecekondu* settlements in defining the fate of politics. *DP*'s approach in relation to the increasing potential vote of *gecekondu* settlements in determining the course of politics had been considered by the party through an integrative political discourse capturing urban and rural poor. The number of *gecekondus* in the periphery of the cities increased during those years. Therefore the loose settlements of *gecekondu* houses in the periphery of the cities drew greater public attention. Public perception in this period mainly signified the transitory character of these settlements like the approaches during Single Party regime. Closer encounters between rural migrants and the "urbanites", who had deeply internalized Republican ideals and been educated in accordance with its doctrines, led to the emergence of a biased approach to *gecekondu* settlements in public discourse. The elitist bias fed by the modernization ideals of Republican Regime took on a more concrete and conservative form during this period.

The urban elites who took critical positions were severely critical of these settlements and *gecekondu* lifestyle. *Gecekondus* were considered as deviant and ugly spaces that ought to disappear from the face of beautiful cities. The definition of *gecekondu* settlements as the "other" of Turkish modernization has taken place within such a context that solidifies the framework of modernist space and lifestyle that have been unattainable to a large extent in reality. This representation of *gecekondu* settlements and *gecekondulus* as the "other" of Turkish modernization in public discourse attribute a false homogeneity to *gecekondu* society and to these settlements.

As mentioned above, the academic works mostly in the disciplines of sociology, political science, urban planning and economy have considered the emergence of *gecekondu* settlements in the periphery of the large cities as an important subject matter during the 1950s and 1960s. The official history of *gecekondu* settlements offered in academic works presupposes a certain discontinuity between the formation of *gecekondus* in the periphery of the cities and the development of city space in general. The presupposition resides in attributing a certain artificial autonomy to these settlements and the people living there. In some of the academic works, the emergence of settlements like *gecekondu* was not linked to the political and economic conditions but rather to the attributes of *gecekondulus* themselves. In this outlook, the causes leading to the emergence of *gecekondu* settlements are limited to the *gecekondulus* themselves, as if their existence is autonomous from the rest of the society. This is the crucial point that will be scrutinized in the context of this study. Academic works in this period capture a belief in the transitory and rural character of these settlements; within this context, the everyday habits and tastes of *gecekondu* families are described in a discourse of degradation so as to solidify the "rural other" of the modernization process, as the author has formerly elaborated in her master thesis.[5]

The second period runs from 1966 to 1980. In 1966, *gecekondu* law no. 775 was promulgated as the earliest legal document denoting the first formal recognition of low-income settlements with its publicly known name, *gecekondu*. This period ends in 1980 when the political and economic structure was totally changed by the intervention of military in democracy. This period witnessed the extreme polarization of politics between right wing and left wing ideologies. The *gecekondu* space, like other places in the cities, was dominated by violence and the struggle between these extreme political groups. This seems to be closely related to strengthening of the image of the *gecekondulus* as a "political target" during this period and the political parties intensified their dealings with *gecekondu* settlements that had triggered the emergence of radical politics in *gecekondu* settlements.

Another important characteristics of this period is the deconstruction of the homogenizing effect of *gecekondu* identity previously imposed on *gecekondu*

---

5 Tok (Demirtaş), Neslihan (1999), (A Critical Approach to Gecekondu Studies in Turkey with Reference to Modernization Theory), Unpublished master thesis, Ankara: Bilkent University. To a large extent, Chapter 2 of this publication is based upon this work of the author. To prevent in this respect misunderstandings, the author wishes to differentiate between her independent academic achievement and a publication of her former thesis-advisor, Erman, Tahire (2001), "The Politics of Squatter Studies in Turkey: Changing Representations of Rural Migrants in the Academic Discourse", *Urban Studies*, 38:7, pp. 983-1002. In this article, Professor Erman, an Associate with the Department of Political Science at Bilkent University, Ankara, has published the core essence of the MA thesis of the author without taking her consent. This constitutes a serious issue since the paper Professor Erman presented together with the author on the Sixth National Social Sciences Conference in Ankara (1999) drew exclusively upon the thesis quoted above.

dwellers as the "rural other" by public discourse. The struggles between radical militant groups and the extreme politicization of *gecekondu* settlements led to the development of awareness about the heterogeneity of *gecekondu* dwellers in hometown, sectarian and ethnic terms. In the public's perception, these differences had been considered as triggering the emergence of serious conflicts particularly in these marginal spaces in comparison to the other neighborhoods in the city. In addition to the political cleavages, the religious divisions became part of the political competition. The sectarian differences and close encounters between *Alevi* and *Sünni*[6] communities in *gecekondu* settlements were articulated with the political struggles between left and right. The conflicts between the leftist and rightist militant groups had found support mainly amongst *Alevi* and

6 *Alevi* people living in Ankara *gecekondus* appear to be descendants of rebellious tribal groups that were religiously affiliated with the Sfavids. See Bruinessen, M. V. (1996), "Kurds, Turks and the *Alevi* Revival in Turkey," *Middle East Report*, p. 7. Their native language is Turkish and they have migrated from Central Anatolian provinces. The practices of *Alevism* greatly differ from the *Sünni* Islamic practices. Prayer (*namaz*), the fast in Ramadan, pilgrimage to Mecca and zakat that are binding duties of *Sünni* Islam are either not practiced by *Alevi* groups or practiced in different ways and times. *Alevis* have their own practices like ceremonial meeting, namely, *Cem*. The ceremony is conducted in the place called *Cemevi* meaning *Cem* House. *Alevis* often built this place as to be used specifically for *Cem* ceremonies. However, there are cases in which big salon of a house such as living room may be used for *Cem* ceremonies. The *Alevis* do not go to *Cemevi* in the way orthodox *Sünnis* go to the mosque. Unlike a mosque or church, *Cemevi* is not a place where one goes for prayer and in order to comfort her/himself through prayer. Contrary to that, in order one to go to *Cemevi*, one need to have a peaceful conscience at first. See Çamuroğlu, R. (2000). *Değişen Koşullarda Alevilik*, İstanbul: Doğan Yayınları, p. 83. Rather than performing the external (*zahir*) demands of Islam *Alevis* claim to live according to the inner (*batin*) demands of religion (Bruinessen, 1996: 7). In that sense to be a good person and development of personal morality come before everything else in *Alevi* belief system and practices. The interpersonal relations within the community; their survival with mutual respect and appreciating good humane characteristics like, tolerance, equality and freedom constitute the backbone of *Alevi* ethical system. Approach to gender differences constitutes another important difference between *Alevi* and *Sünni* belief systems. *Alevi* approach to woman embraces liberal and egalitarian elements; therefore among migrant communities in cities *Alevi* women are most of the time more educated, more occupied and expressive than *Sünni* women. Before the 1950s where *Alevi* and *Sünni* people mostly lived in villages of their own, these identities had not led to any conflict-ridden situation between groups. However, after the 1950s with the intense rural-to-urban migration these identities came into close contact in *gecekondu* settlements constructed in the periphery of the cities.

Sünni communities in *gecekondu* settlements respectively. The public discourse, unlike its emphasis of the "rural other" as a homogenous group having "inferior" cultural qualities in previous decades, had overemphasized the sectarian and ethnic differences among *gecekondu* society in the 1970s to point out the threatening and conflict-ridden qualities of these settlements for national unity.

The third feature of transforming public discourse about *gecekondu* settlements in second period is related to the commercialization of the *gecekondu*. In the early years of rural-to-urban migration, *gecekondu* settlements were mainly built to meet the immediate needs and for direct use of migrants. At that time, *gecekondu* had no or little market value. However, from the 1970s onwards, we can talk about the commercialization of *gecekondu* settlements due to the continuous migratory flow into these settlements and the transformation of *gecekondu* land and property into one of the main sources of economic gain, particularly for the early settled *gecekondu* dwellers. The perception of *gecekondu* settlements as shelters for the poor in the early periods of *gecekondu* formation was accompanied by a belief in the transitory qualities of these settlements. In public debate, this belief had also changed during the 1970s owing to the commercialization of these settlements. In addition, in this period *gecekondus* took on the form of extended neighborhoods constituting large *gecekondu* regions encircling the big cities.

The third period starts in 1980 and covers the years until present. At the beginning of the third period, *gecekondus* had become legalized to a large extent with the implementation of neoliberal policies by the early 1980s and the modernization ideal foreseeing the integration of *gecekondu* settlements with the "modern" urban space had totally failed. The evidence of this can be found in the permanency of these settlements and their continuous growth in various socio-spatial forms.

During this third period, new academic efforts to redefine and understand the changes in *gecekondu* spaces became more widespread. "*Gecekondu*" in its original meaning has become insufficient to define all low-income settlements in the periphery of the cities. On the one hand, there are neighborhoods that were once *gecekondu* settlements but had become legalized to a large extent during this period. They still keep their traditional physical appearance composed of single-story *gecekondu* houses with gardens. On the other hand, there is another type of low-income settlement that has been illegally built after the 1980s with the extremely rapid rural-to-urban migration to large industrial cities such as İstanbul, İzmir, Diyarbakır, and Mersin. These newly emerged low-income settlements reflect a scene of disorganized two or three story houses with an amorphous and unfinished appearance.

In the cities mentioned above, the rural to urban migration has been intensified particularly after the 1980s as a result of forced migration. Therefore the newcomer migrants are mostly Kurds in ethnic origin. The ethnic cleavage seems to add to the political and religious cleavages of low-income settlements from the perspective of public perception. The presence of Kurdish migrants in increasing numbers has triggered the anxiety in public perception about how urban life has been threatened in the 1990s. In fact, a new term "*varoş*" appeared within this context, and has gained an extensive usage after the first half of the 1990s.[7] In relation to that anxiety, the debates regarding *varoş* or its representation in public debates from middle class perspective seem to incorporate, in some cases, racist overtones employing even arbitrary classifications between "White Turks" and "Black Turks", which will be explained in detail in the second chapter. *Varoş*, as used in place or with the concept *gecekondu*, is usually associated with the newly emerging illegal settlements. However, *varoş* is also equated with poor taste and marginal and criminal life styles that are associated with all low-income settlements, new or old, without exception. Public representation has defined all low-income settlements as the breeding ground of criminal activity relating that to the poor living conditions in these spaces, which was actually a consequence of neoliberal economic policies.

Moreover, the electoral success of the conservative Islamist parties in both local and national elections in the 1990s - mainly with the support of low-income settlements- has led to the emergence of a new cleavage; Islamist versus secular.[8] The increasing power and influence of *Sünni* Islam on everyday life and the increasing intensity of tariqat membership in low-income neighborhoods had invoked anxiety among the secularists, mostly the Republican urban elites. Low-income settlements as sheltering these communities were perceived as a threat, in accordance with this general perception. "Urbanites" who perceive secularism as their and nation's defining identity are therefore positioning themselves against "ethnically distinct" and "religiously fundamentalist" *gecekondulus* at the present time.

As briefly explained above, urban middle class perspective in Turkish public discourse had defined its "other" differently over time depending on

---

7 Etöz, Zeliha (2000), "Varoş: Bir İstila Bir Tehdit," *Birikim*. 132.

8 According to many students of Turkish politics, Islamist vs Secular division constitutes a significant explanatory factor in the political spectrum in Turkey. In general, Islamists have been seen as a threat to secular characteristics of the Turkish state. Islamist political parties are not radical movements trying to change the legal and political systems rather, they are arguing for a conservative society based on *Sünni* values. For a detailed analysis of the issue see Ergun Özbudun, 2000 Özbudun, Ergun (2000), *Contemporary Turkish Politics: Challenges to Democratic Consolidation*, Colorado: Lynne Rienner. In general, Islamists have been seen as a threat to secular characteristics of the Turkish state.

the context but with strong reference to *gecekondu* spaces most of the time. This has been done mostly with certain neglect of the interconnectedness between the socio-spatial transformations in low-income settlements and macro political economic strategies by attributing an atomistic quality to the agency of "*gecekondu*" inhabitants. Given that, I would like to analyze in detail some moments through the historical narratives of *gecekondulus* whereby, the acts in the strategic realm, -the realm of politicians, state bureaucrats, urban planners and intellectuals- in defining *gecekondu* space and the tactical everyday acts of *gecekondu* inhabitants, who have used the impediments and loopholes of strategical realm, are dialectically intertwined. In the following section, I will introduce the research questions that reflect the problems of the representation of *gecekondu* settlements in public debate over time via a critical local history of a typical *gecekondu* neighborhood.

1.1.2 Research Objectives

This study intends to discuss two main objectives. First of all, it aims to show the dynamics and patterns by which the strategic interventions in local space and tactical acts of the migrants in producing their locality are closely interconnected. This interconnectedness between the *strategical* and the *tactical* realms will be analyzed mainly through the local spatial history of a typical *gecekondu* neighborhood. In order to analyze this relationship, I have conducted in-depth interviews and observations in Boğaziçi, a neighborhood of Mamak district of Ankara. Mamak is a typical "*gecekondu*" locality which comprises mainly one-story *gecekondu* houses with small gardens that although legalized approximately 20 years ago, nevertheless have kept their traditional physical forms. Boğaziçi neighborhood, the research setting, is an old *gecekondu* neighborhood with a 50 years history. My sample, in that sense, covers inhabitants who have witnessed each of the three periods of *gecekondu* development as explained in the previous section.

Revealing the interconnectedness between strategical and tactical realm will also signify the parallel moments in the spatial transformation of the designated research setting and city space in general. This transformation indicates that the production of *gecekondu* space as a process can only be understood as an integrated part of the general political, social and economic context of Ankara. Therefore the transformation of the neighborhood space from a predominantly natural space[9] in the 1950s to a radically political

9 With reference to the initial settlement period, the respondents mainly described the locality with reference to its natural qualities. Natural factors had determined the course of spatialization mostly during that period.

space in the late 1970s and to an economically and religiously dominated space beginning by the 1980s, gives clues about this interconnectedness and continuity between the macro socio-political developments in Turkey and in the local history of *gecekondu* neighborhoods.

The second main objective of this book is to put forward an alternative local historical reading of the formation of a *gecekondu* space as a response to modernist consideration of *gecekondu* development in Turkey. This alternative historical reading will depend on a narration of the "social spatialization" in the neighborhood. This term, as used by Rob Shields[10] refers to an ongoing social construction of the spatial at the level of social imaginary and representations as well as real interventions in the form of social practices producing a built environment.[11] To put it differently, the concept signifies the process in which social and spatial have continuously defined each other in the locality in close interconnectedness with the general context of urbanization in the city. The analysis of social spatialization, so as to put an alternative history of the locality, has twofold implications.

First of all, the *spontaneity* as the defining and intrinsic quality of *gecekondu* settlements as different from other "more planned" neighborhoods of the city will be made explicit. This spontaneity is not only closely related to the tactical acts of the migrants in relation to the strategical realm, but is also related to certain spatial contingencies emerging in the course of social spatialization. Understanding this spontaneity intrinsic to *gecekondu* settlements makes it possible to reveal the causes behind the biased estimates of strategical planning policies, which are developed in accordance with modernization principles.

Secondly, the examination of social spatialization considering the interconnectedness of tactic and strategy on the vertical level will also highlight the competition between different groups along *hemşehri* and sectarian identities on the horizontal level during the formation of neighborhood space. In other words, the power relations between different groups in the locality have a strong connection with these groups' diverse ways of forming relations with the actors of the strategical realm. The ethnic identities through *hemşehri* and sectarian affiliations directly define the course of spatial production in the locality, since these identities are used by migrants as a means to employ certain tactics in dealing with strategical realm, depending on the nature of identity politics as defined by the

10 See Shields, Rob (1991) *Places on the Margin*, London: Routledge and also Shields (1998) *Lefebvre, Love and Struggle: Spatial Dialectics*, London&New York: Routledge.

11 Shields (1991:31).

strategical realm at that particular moment. Strategical acts have attributed certain values and meanings to these ethnic identities in conjunction with policy interests, which enable migrants to use these identities as tactics in dealing with the strategical realm and other ethnic identities in the locality. In this respect, the nature of competition between these groups in the course of social spatialization in the locality seems to depend strongly on general socio-political context of the city and the country at a particular time. In this sense, the local historical reading of social spatialization following the route mentioned above will also exemplify the nature of encounters between different groups in the neighborhood at different moments of this process. As mentioned above, public representation had either under or overemphasized the heterogeneity of *gecekondu* society. The elite perception either defines *gecekondu* society as a homogenous one by attributing a socio-cultural otherness to the whole *gecekondu* society or exaggerates the threatening qualities of close encounters between conflict ridden identities in *gecekondu* setting to define it as an illegal or criminal space. The narration of the course of social spatialization in the research setting signifies the fact that the *hemşehri*, ethnic and sectarian differences have played important roles as variables that have defined the context of encounters between different groups at certain moments of spatial production. These encounters, however, may sometimes take a strong tone of conflict and struggle or, in other times, may necessitate compromises in the 50 years history of the research setting.

The secondary objective of this study, stemming from the main objectives, is to provide a spatial approach as a methodological and analytical tool to the study of low-income settlements. As the research data illustrates, low-income settlements have strong spontaneous quality based on spatial contingency. The spatial approach allows us to reveal characteristics of *gecekondu* settlements. I draw mainly on the conceptual and theoretical framework as offered in debates on urban (socio)-spatial production in modern societies both at conceptual and methodological levels. In this sense, first of all, I will explain the conceptual framework of the study and define the concepts and the outlook that will be employed in the course of the book. Secondly, I will define the methodological approach of the thesis that attempts to reconcile the local historical reading in a neighborhood depending on the narratives of the respondents and the spatial analysis that accompany and constitute the core of this reading.

1.1.3 Conceptual Framework

The book seeks to critique the exclusionary and inculpative modernist discourse on *gecekondu* space and *gecekondulu* on a number of points as mentioned above. In this process, the *gecekondulu* is defined by public representation either as the main culprit of the mal-urbanization within the "blaming the victim" tradition or as the idle, marginal and passive beneficiaries of Turkish urbanization. In either case, the formation of *gecekondu* space and city space seems to be thought of as separate, thus, concealing or neglecting the paradoxes and impediments of modernist planning activity and the politics of urbanization in Turkey. In conjunction with the research interests, the critical literature on the modernist planning of urban space and daily life constitutes the main inspiration and conceptual framework of the thesis. The critique that Henri Lefebvre[12] poses to the production of space in Western capitalist societies in his groundbreaking book, "The Production of Space" constitutes one of the main inspirations in building up the theoretical framework of this study. The recent debates in critical spatial theory that are enriched not only by the premises but also with the criticisms of Lefebvre's theory construct the conceptual and theoretical skeleton of the book.

The core of the conceptual structure of this study consists of a critical synthesis of Henri Lefebvre's theory on the production of space in modern capitalist societies, -the foundation of critical Marxist geography- supported with Michel De Certeau's[13] theory on the practice of everyday life that aims at bringing an alternative historical reading. The main objective of this section is to show the relevance of the literature with the research interests and to make explicit the conceptual and methodological concerns that played important roles in the research design. In order to accomplish this task, I will define certain concepts that to be employed and refered to. The meaning of (social) space, production of space, social spatialization, spatial triad, strategy, tactic and spontaneity will be discussed at length in connection with their theoretical claims.

1.1.3.1 "Social Spatialization" as a Conceptual Alternative to "Production of Space"

The main objective of Lefebvre's book "The Production of Space" is to show and criticize how state rationalism and power attempts to bring a unified and homogenous society into perfection by producing a peculiar space in accordance with its objectives in modern capitalist societies.[14] In doing this he puts his novel definition of "space" at the center of modern geographical

12 Lefebvre, Henri (1998), *The Production of Space,* Oxford: Blackwell.

13 De Certeau, Michel (1984), *The Practice of Everyday Life,* University of California Press.

14 Lefebvre (1998:281).

thought and points out the vitality of spatial analysis in any social scientific research. In his theory of space, Lefebvre[15] challenges previous definitions of "space". He criticizes Cartesian dualist understanding, where space is defined as an object against subject dominating all senses and bodies. Lefebvre's criticism seems to be mainly inspired by late Heidegger's emerging interest in space alongside time. Heidegger treats questions of spatiality as equally important to those of temporality. As Elden[16] mentions, Heidegger criticizes the understanding of space, as with time, in a narrow, calculative, mathematical sense, divorced from our experience of space in our everyday dealings with the world where we act and react to objects within it in a lived, experiential way, instead of abstracting them in a Cartesian grid of coordinates.

The second main definition of space to which Lefebvre channels criticism is Aristotelian space. This is an empiricist space whose reason of existence is reduced to the classification and naming the evidences of senses rather than having an absolute existence. Lefebvre opposes such a definition of space mainly with reference to the status of space in defining the context of everyday dealings with the world for people.

Thirdly, Lefebvre criticizes Kantian space where it is defined as the apriori realm of consciousness and separated, along with time, from the empirical sphere. The epistemologico-philosophical notion of space in general, fetishizes a mental realm that comes to envelop the social and physical ones, which is challenged by Lefebvre for their neglect of the practical and material qualities of space.[17] According to Lefebvre, recognition of space mainly as a mental realm leads to a gap between the theoretical (mental and social space of philosophers) and practical (space of people dealing with material things). Therefore, theoretical unity should be considered between cosmos (physical nature), mental, logical and formal abstractions and social spaces.[18] He attempts to integrate social practices, perceptions and representations within the definition of space rather than defining space only as an object or physical container of social practices. In this respect, he initiates a third position between positivistic geography dealing with space mainly with reference to its physical qualities and humanistic geography dealing with particular places and the cultural meaning and values attributed to these places lacking a systematic approach. He offers a

---

15 Ibid., p.8.

16 Elden, Stuart (2004), "Between, Marx and Heidegger: Politics, Philosophy and Lefebvre's The Production of Space", *Antipode*, 36 (1), p.92.

17 Lefebvre (1998:5).

18 Lefebvre (1998:11).

combination of materialism, existentialism and phenomenology[19] to come up with a more effective and dynamic definition of space.

Hence Lefebvre's definition of space is quite extensive and inclusive. His definition of space is considered by many to be so inclusive that the whole theory is thought to instigate a fetishization of space over time. In fact, *space* having an equal status with *time* can only be considered as having positional and contingent effects on the context of social relations for some.[20] The reason behind the criticisms directed at Lefebvre is the importance he attributes to the transforming capacities of space in altering the course of social relations in any mode of production. He makes an extended criticism of the modernist production of space as part, parcel and most importantly the determinant and transformer of capitalist relations of production. Particularly, when he contextualizes his analysis in relation to urban space, he signifies the dynamic and dialectic relation of space with the social, emphasizing the defining aspects of spatial contingencies on urban development. Lefebvre's analysis of space and its production in capitalist society points out space's intrinsic capacity to alter the whole context of social relations of production in the following manner:

> Space has a sort of reality of its own, a reality clearly distinct from, yet much like, those assumed in the same global process by commodities, money and capital... space thus produced also serves as a tool of thought and action... in addition to being a means of production, it is also a means of control, and of domination, of power... Is this space an abstract one? Yes, but it is also "real" in the sense in which concrete abstractions such as commodities and money are real. Is it then concrete? Yes, though not in the sense that an object or product is concrete. Is it instrumental? Undoubtedly, but, like knowledge, it extends beyond instrumentality.[21]

---

19 As stated by Andy Merrifield: "Implicated in this project was Lefebvre's own particular brand of Marxism which stressed the importance of everyday life, of alienation and of the writings of the early humanist Marx. Consequently, his project on space does not simply reduce the mental to the material in a 'vulgar' Marxist fashion. For Lefebvre, the realms of perception, symbolism and imagination although distinguishable, are not separable from physical and social space." Merrifield, A. (1993), "Place and Space: A Lefebvrian Reconciliation," *Trans. Br. Inst. Geography*, 18, p.523.

20 See Sayer, Andrew (1985), "The Difference that Space Makes", in Gregory, D. and Urry, J. (eds.), *Social Relations and Spatial Structures*, London: Mac Millan.

21 Lefebvre (1998:26).

His theory on space reflects the power of the production of space as the explicit/implicit manifestation of multifaceted class interests. That is the reason why he uses space with the concept of production as showing the "social" and "power" ingredients embedded in the spatial. He defines "(social) space as a (social) product."[22] Lefebvre's using of "social" in brackets does not signify the fact that space has a detachable social component; on the contrary, he points out, as the first step in his theoretical premise, that "space" and "social" in the modern world are impossible to tell apart. This is the reason why he always uses the concept of space, while dealing with modernity, with the complementary verb "production."[23] The first implication of his theory that space is a social product supposes the disappearance of (physical) natural space with the growing effect of capitalist spatial production. In the capitalist system, nature comes closer than ever to being defeated and destroyed. However, the disappearance of natural space through the creation of certain spatial contingencies may, in some cases, alter the course of the social relations of capitalist production and reproduction.

The second implication of the theory supposes that every mode of production produces a space of its own.[24] To restate it, every power structure, in order to impose its power in an efficient way, produces its own space. In modern times, the space that is produced by capitalist production relations in order to sustain the continuity of the system, is

22 Ibid., p.26.

23 He makes a distinction between "production" and "work" where he defines production in the service of repetition mainly as part of the capitalist system. He attributes "work" all the creativity. Therefore, pre-industrial cities and their spatial qualities are mainly identified within the sphere of work and creativity, but the aspects of modern space are identified with the verb "production" in Lefebvre's terminology. Lefebvre (1998:77). He makes a similar distinction between domination and appropriation of space. According to Lefebvre, domination of space is mainly a quality of capitalist production unlike the appropriation of space. "Domination by technology tends towards non-appropriation- i.e. towards destruction... There is a conflict between domination and appropriation. This conflict takes place in space". Lefebvre (1998:343). Appropriated space is a natural space which serves the needs and possibilities of a group from which it has been appropriated. Lefebvre (1998: 166). Lefebvre sees the resistance capabilities of the powerless in modern societies in their capacity to appropriate space and thus challenge the abstract space produced by capitalist power holders.

24 Lefebvre (1998:32).

called an "abstract space"[25] by Lefebvre. In order to prevent any kind of mass revolt, the "abstract space" of capitalism should conceal relations of exploitation and alienation in everyday life. Therefore, the "abstract space" of capitalism is "buttressed by non-critical (positive) knowledge backed by a frightening capacity for violence and maintained by a bureaucracy which has laid hold of the gains of capitalism" as defined by Lefebvre.[26] The Republican Regime's attempt to create a modernist space seems to include production of such an abstract space which serves the interest of the Republican Regime. It creates a homogenous and modernist physical and social space in the course of nation building as mentioned in the introduction of this chapter.

If space is a product of our knowledge, then it must be expected to reproduce and expound the process of production.[27] The third implication of the theory gives important clues about Lefebvre's notion of space, not only as a social product but also as having a reality of its own, particularly within the context of its capacity to define the reproduction of social relations. Edward Soja[28], a leading interpreter of Lefebvre's theory, develops the term socio-spatial dialectics, particularly with reference to this precept of Lefebvre's "the production of space" refers to and occupies what Soja calls socio-spatial dialectic in which space is conceptualized both as the medium and the outcome of social action, which, according to him, connects social and spatial structures in such a way that the former appears in its concrete form in the latter.[29] Soja's basic premise upon which the "material interpretation of spatiality" is built, reformulate the understanding of the production of space in a Lefebvrian sense. Soja lists the qualities of "concrete spatiality" in the following manner:

---

25 Abstract space "as a product of violence and war is political; instituted by a state, it is institutional. On first inspection it appears homogenous; indeed it serves those forces, which make a tabula rasa of whatever stands in their way, of whatever threatens them- in short of differences. Lefebvre (1998:285). In order to understand Lefebvre's definition of abstract space, there is a need to understand the difference he supposes between domination and appropriation. Abstract space of capitalism dominates rather that appropriates.

26 Ibid., p.52.

27 Ibid., p.36.

28 Soja, Edward (1985), "The Spatiality of Social Life: Towards a Transformative Retheorization" in Gregory, D. and Urry, J. (eds.), *Social Relations and Spatial Structures*, London: Mac Millan.

29 Ibid., p.94.

> 1) Spatiality is a substantiated social product, part of a "second nature" which incorporates as it socializes and transforms both physical and cognitive space; 2) As a social product, spatiality is *simultaneously the medium and the outcome* ... social action and relationship; 3) This spatio-temporal structuration of social life defines how social action and relationship are materially constituted, made concrete; 4) The constitution/concretization process is problematic, filled with contradiction, conflict, and struggle; 5) Conflict and contradiction arise primarily from the duality of produced space as both outcome-embodiment-product and medium-presupposition-producer; 6) Concrete spatiality is thus a competitive arena for both social production and reproduction, for social practices aimed either at maintenance and reinforcement of existing spatiality or at significant restructuring and possible transformation; 7) *The temporality of social life,* from the routines and events of day-to-day activity to the longer-run making of history, *is rooted in spatial contingency in much the same way that spatiality of social life is rooted in temporal/historical contingency;* 8) The materialist interpretation history and the interpretation of spatiality are inseparably intertwined and theoretically concomitant, with *no inherent priorization of one over the other.*[30] (Italics mine)

Soja's reformulation of Lefebvrian understanding of the production of space acknowledges the defining capacity of space in the reproduction of social relations and the effect of spatial contingencies in producing unintended effects. These spatial contingences may alter development schemas in urban context just as historical contingencies may alter the spatial relations of production in such a reformulation.

By introducing such a broad and dynamic definition of space, Lefebvre is attempting to transcend the Marxist dichotomies and binary oppositions between proletariat an bourgeoisie, wages and profit, or productive labor and parasitism. This is done by integrating a third element that is *land* with its determining capacities into two means of production or moments of capitalism, namely *labor* and *capital.*[31] He attempts to go beyond the weaknesses of Marxist formulation of capitalist relations of production by integrating the "production of space" into this formulation. However, the "production of space" in Lefebvrian reformulation emphasizes mainly the

30 Soja (1985:98).

31 Lefebvre (1998:228).

capitalists' power and capacity in producing a totalizing space dominating all other space(s). This reformulation, however, leaves his argument hostage to misinterpretation and a reduction back to established Marxist concepts of production, according to Rob Shields.[32] In fact, Lefebvre's broad definition of space and the capacity of spatial relations in altering the course of social relations should transcend such a homogenous and one-way formulation of power relations of "producing space". Therefore, according to Shields, an important reviewer of Lefebvre's contribution to spatial theory, there emerges a desperate need for a vocabulary to conceptualize the varied production and consumption of varied spaces, places and landscapes. He explains this need to reconceptualize Lefebvre's term of "production of space" in the following manner:

> Lefebvre is referring not only to the empirical disposition of things in the landscape as "space" (the physical aspect) but also attitudes and habitual practices. His metaphoric l'espace might be better understood as the spatialization of social order. In this movement to space, abstract structures such as "culture" become concrete practices and arrangements in space. Social action involves not just a rhythm but also geometry and spacing. Spatialization also captures the processual nature of l'espace that Lefebvre insists is a matter of ongoing activities. *That is, it is not just an achieved order in the built environment, or an ideology, but also an order that is itself always undergoing change from within through actions and innovations of social agents.*[33] (Italics mine)

As mentioned by Shields, unlike the critics of Lefebvre who point out the weaknesses of his Marxist framework, Lefebvre appreciates the actions and innovations of social agents within the context of his critique to structuralism. Particularly when he refers to urban space and its modernist production and planning, he signifies the irreconcilability between the plans of the center and the spatial practices of the "users" in daily life. The power of the "users" is implicit in their ability to appropriate alternative spaces for their needs incongruence with modernist attempt of spatial production, according to Lefebvre. This has a close connection with the creativity and dynamism he attributes to the social agency and the capacity of the "users" in altering the course of spatial production. However, this aspect of his theory remains

32 See Shields, R. (1998), *Lefebvre, Love and Struggle: Spatial Dialectics*, London&New York: Routledge, p.154.

33 Ibid., p.155.

open to misreading, owing to his conceptualization of capitalist space as a totalizing space assimilating all *"other"* spatial experiences, as defined via the term "production of space". The main reason behind Shields' attempt to put the term "social spatialization" in place of the "production of space" is to overcome such a misleading reading of Lefebvre's theory. It attempts to overcome the internal conceptual contradictions of Lefebvre's terminology in order to strengthen Lefebvre's critique of the modernist, rationalist system of capitalist production via the novel understanding of the dialectical relation of the social and spatial. Social spatialization, in this respect, not only designates the physical environment but the process in which social practices and social imaginary and spatial constructions are in continuous interplay.

Shields[34] defines "social spatialization" as the ongoing and processual social construction of the spatial at the level of the social imaginary (collective mythologies and presuppositions) as well as interventions in the landscape (for example, built environment). Social spatialization as a process also includes contingency and spontaneity as it emerges from the spatial determination of the social relations, thus constituting the core of Lefebvre's novel definition of space as detailed in the following pages. The term "social spatialization" is preferred in place of "production of space" in most parts of this book with reference to the discussion above. Lefebvre's formulation of spatial production in interconnection with the social signifies a variety of power positions and interests, rather than referring to a homogenous realm with a unilateral power structure. In relation to that, I will introduce Lefebvre's spatial triad, which will be referred to frequently throughout this study.

1.1.3.2 Spatial Triad

Lefebvre's main criticism to modernist urban planning is made explicit through his formulation of a spatial triad. According to Lefebvre, every society has its peculiar spatial code and this spatial code can be defined with a spatial triad. In this triad, space is defined as an outcome of the dialectic interplay between the three moments of social spatialization or acts of social agency. The first component of the triad is *spatial practice,* that is society's space. Society produces this slowly and surely as it masters and appropriates it in everyday life. That can be defined as an externalized material or built environment. This is the perceived space in which there is a close connection between "daily reality" (daily routine) and "urban reality" (the routes

34 See Shields, Rob (1991), *Places on the Margin,* London: Routledge, p. 31.

and networks which link up the places set aside for work, private life and leisure.[35] The reproduction of social relations by means of real interruptions and activities in space take place in this aspect of the spatial triad.

The second component of the triad is the *representations of space* (conceived space), that is, the space of planners, urbanists, technocratic subdividers and social engineers. This is the part where a conceptual model, used to direct practice, is imposed by modernist or capitalist agencies. "This space comprises the various arcane signs, jargons, codifications, objectified representations used and produced by these agents."[36] The second part of the triad constitutes the core of Lefebvre's criticism posed by modernist planning strategies and their incompatibly with the "lived space", or the space of everyday life. The problem under capitalism is the primacy given to the determining capacity of the conceived part of the spatial triad.

The third component of the triad is *representational space* (lived space) that is space directly lived through its associated images and symbols, and hence, the space of "inhabitants" and "users", and also some artists. It is the sphere where the lived social relations of users in relation to the environment are conceptualized.[37] This is the space of "users" or "inhabitants" and it is where Lefebvre sees the source of potential challenge to the existing rationalist system which has also the potential to create difference.

Lefebvre offers such a spatial triad to explain the production of space in a capitalist society and the power relations between different agents in such a framework. This triad integrates a dynamism and expansion to the dialectics of the social and spatial. It avoids, at the same time, a degeneration of the discussion of the dualistic socio-spatial dialectic into a reductionist and unproductive debate over the primacy of the social over space or vice-versa.[38] According to Lefebvre, these three aspects -perceived, conceived and lived- go together; and there are no antagonisms, oppositions and contrasts between these three aspects of the triad, but rather, echoes, repercussions and mirror effects.

The reason why Lefebvre sees no antagonisms between these three aspects may depend on his overemphasis of the power of the capitalist system to create a totalizing space where the "users" in the representational space have little maneuvering ability to affect the agency in the other two realms,

---

35 See Lefebvre (1998: 37, 50).

36 Merrifield (1993:523).

37 Lefebvre (1998:38, 39); See also Gottdiener, Mark (1993), "A Marx of Our Time: Henri Lefebvre and the Production of Space," *Sociological Theory*, 11 (2), p.131.

38 Shields (1991:56).

the representations of space and spatial practice. In this respect, Lefebvre defines the representational space as the dominated and hence passively experienced space, the space of the "users" rather than the "producers", which the imagination seeks to change and appropriate.[39] This is the most criticized and problematic part of Lefebvre's theory in general, but also for the objectives of the thesis. To exemplify the potential of *gecekondu* inhabitants to produce their own space through the use of loopholes in the modernist planning and policies necessitates, conceptually, an opening in this part of Lefebvre's theory. In order to overcome this theoretical impediment, De Certeau's distinction between strategy and tactic will be synthesized with Lefebvre's triad with a critical rereading of both theories and will constitute the subject matter of the following sections. Shields also directs his main criticism of Lefebvre at that point. As indicated by Lefebvre, the capitalist system produces an abstract space that is imposed as a hegemonic and totalizing space, mentioned in the previous section. This leads to much tortuous writing and a tendency to reduce the contemporary social spatialization to a capitalist strategy. Yet, a close reading of Lefebvre's triad also signifies the potential he sees, mainly in the realm of representational space. As stated by him, the producers of space always acted in accordance with a representation -very difficult to challenge- while users, on the other hand, "passively experienced whatever was imposed upon them in as much as it was more or less thoroughly inserted into or justified by their representational space."[40] In this statement, Lefebvre makes explicit the necessary condition behind users' acceptance of what is imposed upon them. If the space of the "producers" (representations of space) is in great disharmony with the representational space of the "users", than the users may act to create a maneuvering space for themselves, which is again limited by the knowledge and power of the "producers".[41] Still, the existence of a representational space having a certain capacity to create a *differential space* poses one of the most important challenges to the capitalist system's production of a unified and homogenous space.

To sum up, Lefebvre's triad is important in his critique of structuralism in general and structural Marxism in particular. He opposes all dualities and a dichotomist reading of power relations and structures by exhibiting a triad of spatial production. The creativity of the social agency is appreciated in his theory on space yet it is not easy to get this fact out because he often states the power of capitalist system in creating an abstract space that is overemphasized

39 Lefebvre (1998:39).

40 Ibid., p.43, 44

41 Ibid., p.50.

and portrayed as imposing its representations on the "users". The reader may have a sense throughout his book of the impossibilities of the "powerless" or "users" to resist, or about their lack of potential to create a resistant space. As Shields mentions, despite this difficulty, "The Production of Space" is "devoted to developing a radical phenomenology of space as a humanistic basis from which to launch a critique of the denial of individual and community's "right to space" under the abstract spatialization embodied in the capitalism and technocratic knowledge structures of the state."[42] In this respect, the theoretical elegance of his book offers wide-ranging conceptual and methodological tools for criticizing modernist planning and the potential of representational spaces in creating a spontaneity and difference, even by means of the structural weaknesses of rational and bureaucratic planning. This constitutes the main reason for such a detailed reading of Lefebvre's critique of the modernist attempt at producing space in relation to the interests of this study.

1.1.3.3 Spatial Spontaneity

There is a strong possibility of spontaneity in every holistic urban planning attempt even in cases where the agency of the conceived realm is powerful enough to impose its space. The sources of spontaneity in Lefebvre's triad lie in the components of "representational space". "Representational space" of Lefebvre's triad is;

> Fully "lived" space (*l'espace vécu*)... It is an essential terrain of struggle on the way to realizing ourselves as "total persons"... This sphere offers complex re-coded and even decoded versions of lived spatializations, veiled criticism of dominant social orders and of the categories of social thought often expressed in aesthet ic terms as symbolic resistance... Also included in this aspect are clandestine and underground spatial practices, which suggest and prompt alternative (revolutionary) restructurings of institutionalized discourses of space and new modes of spatial praxis, such as squatters, illegal aliens, and third world slum dwellers, who fashion a spatial presence and practice outside the norms of prevailing (enforced) social spatialization.[43]

As mentioned in the above anecdote, the component of "space of representation" or "representational space" in Lefebvre's triad unlike the

---

42 Shields (1998:146).

43 Ibid., p.164.

criticisms posed, offers an opening for possible resistances and spontaneities within the represented and practiced space of the hegemonic power. This potential of spontaneity in the production of capitalist space is signified a number of times in his book, "The Production of Space". He argues that the growth of the forces of production does not directly in a causal fashion lead to a particular space or time. Mediation and mediators, so as the action of groups, the factors within knowledge, within ideology, within the domain of representations, in short, the "human agency" has to be taken into consideration in order to understand this process.[44] Soja's interpretation and reformulation of socio-spatial dialectic as mainly inspired by Lefebvre foresee such a potential of spontaneity in the production of space. The spontaneity emerged outside the context of modernist's representations of space is both due to the inherent qualities of space and its production and also the unintended consequences of the spatial acts of human agency. In that respect, Soja emphasizes the role of space as with time in the emergence of contingencies. Soja[45] proposes that the temporality of social life, from the routines and events of day-to-day activity to the longer-run making of history, is rooted in spatial contingency in much the same way that spatiality of social life is rooted in temporal/historical contingency. This is mainly related to the peculiar qualities of space as produced in modern society. As maintained by Lefebvre, space as a social product is not only the outcome of the social action but it has a peculiarity differentiating it from other products of capitalism, which is its capacity to produce and reproduce social relations while being produced. This premise brings the contingency to the forefront of the theoretical debate that aims to articulate new geographical approaches with critical social theory. In this respect, Derek Gregory[46] suggests that the centrality of space in human affairs could be translated into and/or articulated with the premises of structuration theory. In structuration theory, language of agency and the bounded contingency of practical life are reconceptualized through spatial structures.

Lefebvre's reference to the spatial spontaneity and contingency of everyday life in modernity owes to Nietszche's conception of everyday life. Nietzschean definition of the modern world as an "assertion of life and lived against political and economic processes" seemed to inspire

44 Lefebvre (1998:77).

45 See Soja, Edward (1985), "The Spatiality of Social Life: Towards a Transformative Retheorization" in Gregory, D. and Urry, J. (eds.), *Social Relations and Spatial Structures*, London: Mac Millan, p.94.

46 Gregory, Derek (1981), "Human Agency and Human Geography," *Transactions of the Institute of British Geographers*, 6(1), pp.1-18.

Lefebvre's integration of the "lived" as the most important aspect of the spatial triad acknowledging potential of users' agency to create differential spaces as resistant spaces.[47] By accepting the importance of reversibility and instability in everyday life, Lefebvre provides an extended place in his theory to the spontaneity embedded in these appropriated spaces. Lefebvre refers to shantytowns in explicating spatial spontaneity. Yet, the emphasis of the hegemonic power of the center is made explicit vis-à-vis the resisting power of these spaces in his theory. As he states, the differences (squatters, shanty towns) endure on the margins of the homogenized realm (modern urban space) either in the form of resistances or in the form of externalities (lateral, heterotopical, heterological). However, for Lefebvre, sooner or later, the existing center and the forces of homogenization must seek to absorb all such differences and they will succeed if differential spaces merely maintain a defensive posture rather than engaging in counter action.[48] As will be criticized in the following sections in a more explicit manner, the "center" or the sphere of "representations of space" is defined with an overall mission of homogenization and production of the "abstract space" of capitalism as defined by Lefebvre. However, it is important to note that, in some cases the system to define and legitimize its "space" may also be willing to keep its "other" as in the form of differential spaces.

Lefebvre's perception regarding the irreconcilability between the space of producers and users defines the abstract space of capitalism almost as pseudo space. In this respect, spontaneous architecture and planning in shantytowns prove greatly superior to the organization of space by specialists, according to Lefebvre[49], who gives Oscar Niemeyer's Brasilia as an example to such planned spaces into which a faithfully technocratic and state bureaucratic society has been projected and he states that there is an almost self-consciously comic aspect to the process.[50] Brasilia carries important similarities to the rationality implicit in the construction of Ankara. The irreconcilability of the "perceived" and the "lived" in the course of the planning and construction of Ankara will be discussed in detail in the third chapter taking on mainly Lefebvrian conceptual framework.

The emergence of the *gecekondu* settlements in the heart of a rational implication of a westernization project, Ankara city, even in the early years of city's establishment can be considered as a good example to the potential of spatial spontaneity embedded in all modernist attempt of producing space.

47 Elden (2004:87).

48 Lefebvre (1998:373).

49 Ibid., p.347.

50 Ibid., p.313.

Such spontaneous spatial formations manifest a social life far more intense than the bourgeois districts of the cities and according to Lefebvre reflect an appropriation of a remarkably high order. This approach is paraphrased by Shields in the following manner:

> Slums, barrios and favelas are seen by Lefebvre as localized "reappropriations" of space that may furnish examples of such "representational spaces" or "spaces of representation" by which certain sites are removed or severed from the governing spatialization and returned to the realm of "communitas". These are prophetic, temporary autonomous zones. Lefebvre differentiates the popular appropriation of space from the dominated space of the nation state or of the capitalist city. The latter is the site of hegemonic forces of capital, the former site of possible emergent spatial revolutions.[51]

The attribution by Lefebvre to these spontaneous spaces the potential of spatial revolution against the abstract space of capitalism can also be acknowledged as his emotional appeal to Marx's conception of alienation. In his criticism of everyday life, he signifies the alienation by referring to the fragmented quality of individual life.[52] Similar approaches to the study of the spontaneous spaces in third world cities that have attributed these spaces a potential of resistance against the homogenizing power of the center are strongly criticized as romanticizing the problems of squatter settlements, and neglecting the systematic problems in political, economic and social terms. These approaches are also criticized for the fact that they attribute positive and transformative potential to the spontaneous settlements. However, looking at it from another perspective, to understand and explain the socio-spatial dynamics behind the emergence of these spontaneous spaces in which the artificial aspects of modernist planning are also made explicit would greatly contribute to develop a more inclusive and democratic approach in urban governance. Since the agency of residents in squatter type of settlements seems to be reasonably effective in defining the construction of these spaces when compared to other districts in the city, one needs to have access to local knowledge explicating the socio-spatial dynamics between groups.

Particularly by the 1980s, with the increasing effect of neoliberal policies in Turkey and in many countries, the spontaneity/contingency issue

---

51 Shields (1998:165).

52 Ibid., p.73.

has centralized recent debates in geographical theory on urban space. Groth&Corjin[53] in a recent case study on the reappropriation of urban residual spaces[54] in everyday life propose an alternative local planning agenda by civil or informal actors coming from outside the official, institutionalized domain of urban planning and urban politics. According to the authors, the three cases from Helsinki, Berlin and Brussels constitute examples of spaces which do not speak the traditional language of planning, thus allowing a "spontaneous urbanity" to arise which has been open to constant change, flexible, almost personalized transformation of space.[55] Such studies that emphasize the spontaneous appropriation of certain spaces as an alternative to the central planning attempts brings two closely related themes into the forefront of spatial theory: the peculiarities of space as capturing the potential of creating contingency more than the other commodities of capitalism, and the agency of the "weak", or "users" in Lefebvrian sense, in playing with such contingency. Particularly, the debates on the neoliberal reorganization of urban space recently rely on this aspect of spatialization.

The formation and transformation of urban space have been predominantly defined by free market mechanisms that have been coupled with neoliberal opening of urban space to national and international investment as mentioned by such debates. However, as a spatially embedded commodity, real estate embodies a crucial paradox.[56] Despite the fact that real estate has always attracted a range of investors-from the small-scale speculator to the largest insurance companies-, the shifting values in urban space can easily cause real estate in certain locations to acquire uneven and fluctuating values. Such uneven shifts in real estate market value stem from the inflexible nature of real estate, which resists frequent modifications as a commodity. "These qualities make the commodity of real estate very sensitive to devolarization, espe cially in contrast to machinery and other forms of fixed capital."[57] Besides, the distinction between use and exchange value of real estate may cause a problem if one considers its inherent resistance to modification. The peculiar

---

53 Groth, John and Corjin, Eric (2005), "Reclaiming Urbanity: Indeterminate Spaces, Informal Actors and Urban Agenda Setting," *Urban Studies*, 42(3), p.506.

54 "Urban residual spaces" refer to abandoned industrial areas- i.e. interstitial sites that are weak in spatial terms may, due to their indeterminate character provide opportunities for new, transitional reappropriations. See Groth and Corjin (2005:506).

55 Ibid., p.509.

56 See Weber, Rachel (2002), "Extracting Value from the City: Neoliberalism and Urban Redevelopment" in Brenner, Neil and Theodore, Nik (eds.), *Spaces of Neoliberalism*, Blackwell Publishing, p.174.

57 Ibid., p.174.

qualities of real estate seem to trigger conflict and competition between different interest groups on urban space. Those people with emotional attachment to place and those without such attachments may stand against each other in this competition.[58] As will be made explicit in the empirical chapters, the neoliberal policies on *gecekondu* settlements have also led to such contingent results in local context in relation to the inherent qualities of space. With the legalization of most classical *gecekondu* settlements as a consequence of neoliberal policies, some *gecekondu* settlements evaluated as favorable by large-scale investors were opened to competitive market mechanisms. On the other hand, some *gecekondu* settlements, due to their low market value in congruence with their location vis-à-vis favorable spaces, experienced this apartmentalization process to a very limited extent. Furthermore, the interest clash of local groups in these spontaneous spaces may also bring about unintended and contingent spatial developments in such a neoliberal context. The spatial practice of local actors and their attribution of use and exchange value to certain places in *gecekondu* neighborhoods when combined with the asymmetrical power relations in localities contribute to the spontaneity of these neighborhoods. In this sense, the spatial acts of the inhabitants of *gecekondu* settlements gain considerable importance when the potential of spontaneity embedded in these neighborhoods is taken into consideration.

1.1.3.4. Strategy versus Tactic

Due to his understanding of the creative and transformative agency of local actors regarding their potential to create spontaneity, Michel De Certeau's[59] critique of everyday life and his concepts are very relevant for this study. De Certeau has also built his arguments on Lefebvre's general critique capitalist abstract space. In his book "The Practice of Everyday Life" Michel De Certeau brings a similar criticism to everyday life as organized by capitalism and the domination of modernist strategic acts in defining a totalizing space for the "consumers" with Lefebvre (whom he credits as the inspiration for the criticism) with a different vocabulary than Lefebvre. The rationalization of the city, in a very similar fashion to Lefebvre's critique, seems to be mystified by the strategic discourse, according to De Certeau. He further indicates[60] that the city founded by utopian and urbanist discourse is defined by the possibility of a threefold operation.

58 Ibid., p.172.

59 See De Certeau, Michel (1984), *The Practice of Everyday Life*, University of California Press.

60 Ibid., p.94.

First of all, the strtegy should produce its own space in which rational organization must repress all the physical, mental and political pollutions that would comprise it. Secondly, univocal scientific strategies must replace the tactics of users who take advantage of "opportunities". Finally, the universal, anonymous subject which is the city itself should be created and thereby it gradually becomes possible to attribute to it all the functions and predicates that were previously scattered and assigned to many other different real subjects.

However, as De Certeau indicated, modernization praxis and paradigm mainly lead to the *cul-de-sac* (dead end) of the functionalist organization, which, by "privileging progress, (i.e., time), causes the condition of its own possibility- space itself- to be forgotten; space thus becomes a blind spot in a scientific and political technology".[61] This inherent inability of the modernization ideal to realize itself leads to a dualism and a distinction between "strategy" and "tactic" as defined by De Certeau.

The distinction between strategy and tactic is the near correspondence of the distinction as supposed by Lefebvre between the realm of "producers" and "users". As stated by De Certeau, the possibility of any strategic act to realize itself, there needs to be a defined and planned space. Lefebvre also refers to the realm of strategy at some point of his discussions on the production of space where he signifies the close and necessary relation between strategy and occupying a certain space. The goal of any strategy, for him, is to occupy space by the varied means of politics and war.[62] The agency of "representations of space", or its correspondence in De Certeau's terminology, "strategic realm" is able to produce, tabulate, and impose spaces.[63] The actors who take part in the macro planning activity (urban planners, bureaucrats, technocrats, social engineers, academics) who have the ability to conceptualize space in the Lefebvrian sense have the economic and political power to develop strategies. As in management, every strategic rationalization seeks first of all to distinguish its own place, that is, the place of its own power and will, from an environment or in other words from the context of actually existing social relations.[64]

On the other hand, the space of the tactic is the space of the other or "weak", it does not, therefore, "have the options of planning a general strategy and viewing the adversary as a whole within a district, visible, and

61 Ibid., p.95.

62 See Lefebvre (1998:366).

63 See De Certeau (1984:30).

64 Ibid., p.36.

objectifiable space."[65] The tactics of the weak can only use, manipulate, and divert these spaces of power. Tactics are procedures that gain validity and success mostly in relation to their efficient usage of time and their ability in turning an instant intervention into a favorable situation where the rapidity of the movements may change the organization of a space, which paves the way for the successful application of a tactic.[66] De Certeau places a relatively more noticeable emphasis when compared to Lefebvre on the autonomy or tactical power of the "weak" to alter the space of strategy by employing the very tools of the strategical realm. The common position as held by Lefebvre and De Certeau is that they have transformed the concept of space and "taken it beyond the formalistic axioms of geometry, transforming the ephemeral character of spatialization into a dynamic enigma".[67]

The distinction between strategy and tactic by De Certeau is conceptually to have the resources to define the acts of *gecekondulus* in the process of diverting the modernist space of strategy, which is by its nature open to exploitation and diversion due to its irreconcilability with the actually existing social relations. Gülsüm Baydar[68] first used the conceptual framework of De Certeau in relation to *gecekondu* dwellers. As stated by her,[69] although *gecekondulus* lack necessary power to "strategically" utilize and manipulate local political institutions in their service, they "display ingenious tactics" in which they "turn events into opportunities and make use of the strong". In a sense, migrants "lend a political dimension to everyday practices" and they protect early *gecekondu* dwellings by using "tactical operations against the relentless strategies of city fathers."[70]

De Certeau also gives examples from the squatter type of formations in his discussion of the tactical realm. He signifies the potential of political agency with a stronger emphasis than Lefebvre referring to the maneuver capacity of the users within the strategical space and their potential of diverting the space. Even so, he defines tactical acts mainly within the realm of consumption rather than production. In a number of points throughout his discussion, De Certeau gives examples from the migrant dwellings in

---

65 Ibid., p.37.

66 Ibid., p.38.

67 Van Loon, Joost (2002), "Social Spatialization and Everyday Life," *Space and Culture*, 5 (2), p. 90.

68 Baydar, Gülsüm (1997), "Silent Interruptions: Urban Encounters with Rural Turkey." in Bozdoğan, Sibel and Kasaba, Reşat (eds.), *Rethinking the Modern Project in Turkey*, Seattle: Washington University Press, pp.192-211.

69 Ibid., p.203.

70 Ibid., p.203.

France, referring to these people's consumption and diversion of city space in accordance with their cultural prerequisites:

> Thus a North African living in Paris of Roubaix (France) insinuates into the system imposed on him by the construction of a low-income housing development or of the French language the ways of "dwelling" (in a house or in a language) peculiar to his native Kabylia. He superimposes them, by that combination, creates for himself a space in which he can find ways of using the constraining order of the place or of the language. Without leaving the place where he has no chance but to live and which lays down its law for him, he establishes within it a degree of plurality and creativity. By an art of being in between, he draws unexpected results from his situation.[71]

He gives the above example for the tactic of the powerless and their potential for creating certain spaces of their own. However, De Certeau defines these dwellings mainly in the context of time-sensitive tactical acts of consuming the dominant space rather than defining these dwellings as more permanent socio-spatial forms that have some potential to influence the strategical realm. If we are to take and employ De Certeauan theoretical framework to understand and conceptualize social spatialization in *gecekondu* settlements, his distinction between strategy and tactic can be criticized on a number of points with an iterative consideration of empirical feedback. In this respect, the critical and merged reading of Lefebvre's theory on production of space and De Certeau's distinction between strategy and tactic will be employed throughout the book with regard to the empirical findings by critically analyzing these two theories.

However, before the empirical data is evaluated, certain criticisms should be brought to both De Certeau and Lefebvre in terms of their conception of power relations and the dualism they cannot avoid, with the supposition of a clear distinction between the strategical and tactical realms with such a power concept.

The supposition of dualism between the tactical and the strategical realm has been defined with a non-dynamic understanding of power relations. Although one of De Certeau's main objective is to point out the potential moments in an alternative reading of history when the tactics of the "weak" divert the dominating space in various ways, his overemphasis on the power of the strategical realm to impose its hegemony over the "users"

71 De Certeau (1984:30).

or "consumers" seems to sabotage this objective to a certain extent. The limits of the tactical realm are so often emphasized by De Certeau that even the ways in which the strategical and tactical realm interact in a practical context seem to be neglected. Lacking a place of its own, lacking a view of the whole, limited by blindness, limited by the possibilities of the moment, a tactic is determined by the absence of power just as a strategy is organized by the postulation of power, according to De Certeau.[72]

The reciprocal relation between the formation of strategy and the employment of tactics appears to be neglected in the perfect and homogenous of this strategical realm. Different definitions of power make it possible to understand the dynamism and reciprocity between the strategical and tactical realm. Ewick and Silbey[73] conducted a qualitative research that depended on respondents' stories about their dealings and encounters with law. Individuals employed the opportunities of the strategical realm to develop resistant acts. The aim of this study was to analyze the individual narratives of resistance. Ewick and Silbey give direct references to the terminology of De Certeau in their study of these narratives. Depending on the case they are studying, they attempt to transcend the limitations of De Certeau's theory and offer new ways of looking at the relationship between the tactical and the strategical realms. According to the authors, acknowledgment of power is not a thing to possess, but a probabilistic social relationship, which helps one to see the interconnections between the strategical and tactical realm. Power is a series of transactions whose consequences are contingent upon the contribution of all parties, both those who achieve greater power (superordinate) and those who achieve less power (subordinate). This is a relational and processive understanding of power. In this respect, there needs to be reciprocity and close alliance between the institutional structures of the strategical realm and resistance. Resistance, as much as power, is contingent upon the structural resources available to the relational participants.[74] Within such a framework, they define certain patterns of acts of resistance as individuals deal with the law. They name these patterns as the "typical reversals of conventional features" of structural relations. As mentioned above, the tactical realm is mainly operationalized by the efficient usage of time. However, Ewick and Silbey[75] go beyond the possible patterns of resistant acts by introducing tactics that depend on

72 De Certeau (1984:38).

73 Ewick, P. and Silbey, S. (2003), "Narrating Social Structure: Stories of Resistance to Legal Authority," *American Journal of Sociology*, 108(6), pp.1328-1372.

74 Ewick and Silbey (2003:1335).

75 Ibid., pp.1350, 1351.

different sources, like "masquerade", that is, playing with social roles; "rule literalness", that is, playing with rules; "foot dragging", that is, playing with time; and "colonizing space", that is the tactical act of appropriating and employing space deviating from the context of strategical space.

The intentional change of social roles in order to deal with law is "masquerade". The tactic of "rule literalness" depends on two sources as stated by the authors. First, the possibility of resistance may emerge from the incompleteness and gap in any rule system that provides opportunities for the "weak" to exploit. This incompleteness may emerge from a lacuna in the system of rules, by virtue of not being governed or defined.[76] The second source may involve subverting the purpose of rule by rigidly observing it. "Foot dragging" is the resister's playing with the rationalized time of the modernist strategy that defines and controls behavior in terms of time.[77] Fourthly "colonizing space" produces spaces of resistance.

With such a novel typology of tactical acts, the limitations of tactic as defined by De Certeau seem to be transcended. Ian Burkitt[78] brings a similar criticism to De Certeau's conception of everyday life. According to Burkitt, De Certeau's definition of tactics as temporary, dispersed and lacking a colonized space should be rethought. The practices tactical in nature (unofficial) do have their spaces and, moreover, the way time and space is combined in such practices differentiate them from official practices.[79] In this understanding, independent from the official or unofficial nature of some practices within everyday life, some practices are fixed more in geographical space and relatively frozen in time, while other more fleeting experiences are quick to pass and do not have such a materialization in geographical space.[80]

To conclude, both Lefebvre and De Certeau seem to accept an asymmetrical power relation as an advantage for the agency of the conceived realm or strategical realm. Lefebvre acknowledges the repercussions between the three components of the triad. According to Lefebvre, the study of space must account for both "representational spaces" (lived) and "representations of space" (conceived) but above all, for their interrelationships and their links with "spatial practice."[81] However, he also overemphasizes the power

---

76 Ibid., p.1353.

77 Ibid., p.1359.

78 Burkitt, Ian (2004), "The Time and Space of Everyday Life," *Cultural Studies* 18(2/3), pp.211-227.

79 Ibid., p.216.

80 Ibid., p.224.

81 Lefebvre (1998:116).

of social engineering projects to create a totalizing and homogenizing space. De Certeau leaves a relatively more extended and flexible place for a tactical act's capacity to divert space from within the system, although the potential and ability of users to create their own space is, to a large extent, neglected because of the failure to identify the "weak" mainly with the act of consumption rather than the act of producing resistant and permanent space(s).

As will be pointed out by the findings of the research in the following chapters, a critical reconsideration of the strategical and tactical realm is necessary. First of all, there is a need to reconsider the theoretical dualism between strategy and tactic because evidence from the research implies their interconnected nature. Within the context of this interconnectedness, it will be argued that not only the acts in the tactical realm need to taken into consideration the parameters of the strategical realm, but also the strategy defines itself and its "other" by taking into account the nature of the incidents in the tactical realm.

Secondly, the diversity and dynamism of power positions should be taken into account within each realm. The diverse and reciprocal ways of actors' dealing or relating with the actors in their or other realms of the spatial triad should be considered. This examination on the empirical level would also bring up the need for new concepts to define acts of *gecekondulus* vis-à-vis the actors in the realm of strategy on the vertical level or the "other" groups in the realm of tactic on the horizontal level. Since the main motivation behind the research is to understand the nature of social spatialization in the neighborhood rather than the power relations between different groups based on *hemşehri*, sectarian or ethnic lines, this examination will be made by showing certain examples to the shifting nature of the relations between the groups, with reference to the critical moments in the course of spatial history. Such a dynamic model of social spatialization would help to bring a dynamic approach to shifting group identities and change the boundaries between groups, depending on the context of the relationship at different moments of social spatialization.

Within the context of a reconsideration of the theoretical frameworks offered by Lefebvre and De Certeau, certain empirical cases will play a vital contributory role. Therefore, in empirical cases I will draw attention to local groups who have been positioned within the tactical realm at certain moments of social spatialization as totally lacking relations with the actors in the strategical realm. Their developed socio-political relations with the actors in the strategical realm on the vertical level have brought about the gain in power of these groups over time. This has turned their actions into a near strategy as a consequence of this shift in power relations.

It will also be exemplified through the use of empirical cases that tactical acts of the powerless, in their attempt to create a space for themselves, may influence the planning acts of the strategical realm in the coming decades, in spite of the strong definition of *gecekondu* settlements as the "other" of the modernist space. The empirical cases will also pose examples by which the homogenous definition of each realm is challenged. The changing relations of power in the tactical realm between *hemşehri*, sectarian and ethnic groups depending on the strategical context and the spontaneous and contingent nature of social spatialization will also be discussed within the context of such theoretical reconsideration.

The study of tactical forms of resistance is generally criticized by some scholars in two main respects, as defined by Ewick and Sibley.[82] First, the purpose of these resistant acts can never be clearly defined; therefore it is not always clear how the practitioners of such acts interpret what they are doing. In this respect, there is a danger of attributing a greater agency and an over-developed oppositional consciousness to these acts.

The second critique supposes that the focus on everyday resistance abandons a vision of transformative politics by a preoccupation with what appears to be individual, often momentarily trivial acts.[83] In conjunction with these criticisms, the authors point out the significance of tackling the traces of the reciprocal relations between resistance and the realm of power, where the resistance does not seize upon lapses of power so much as relies on the persistence of and familiarity with a particular social organization. While keeping such a conceptual framework in mind, the reading of the local historical dynamics of social spatialization in a *gecekondu* neighborhood will be achieved mainly by analyzing tactical acts as an embedded and connected part of the system rather than a marginal and ineffective part of the city. The acts of migrants in appropriation of *gecekondu* lands, the construction of their *gecekondus*, the claiming of certain services like water, electricity, transportation, the construction of religious spaces and associations that affect the course of spatial development, are all considered within the context of the analysis. Rather than romanticizing such acts, as has also been pointed out by Baydar,[84] it is important to understand the intertwined nature of tactic and strategy in compliance with the research interests.

82 See Ewick and Silbey (2003:1330).

83 Ibid., p. 1330.

84 Baydar (1997:208).

1.2 Methodological Framework of the Study

1.2.1 Definition of the Research Setting

A large segment of the designated research setting is included within the official boundaries of the Boğaziçi neighborhood that is located on the edge of Samsun Highway, the main highway that connects Ankara with other central Anatolian towns and cities. It is here where the rural-to-urban migration to Ankara had mainly taken place and the area extends towards the Nato Highway. The research had been chiefly conducted in the Boğaziçi neighborhood, including the places on the boundary between Boğaziçi and adjacent neighborhoods. Some cases are also used from the places that are close to Boğaziçi in these adjacent neighborhoods, namely Akşemsettin, Duralialıç and Fahri Korutürk. Since the main objective of the research is to write a spatial history of the locality rather than of a specific neighborhood, it is not restricted to Boğaziçi only.

The locality is considered officially as part of the Nato Highway region and constitutes an important component of the Mamak municipal district. The reasons behind selecting the Mamak *gecekondu* region are closely related to the certain pecularities and commonalities of Mamak vis-à-vis other *gecekondu* districts. Mamak is one of the largest squatter districts in Ankara and is commonly recognized as the oldest and most typical squatter district in Turkey. It is composed of fifty-six neighborhoods and six villages, each with an approximate population of ten thousand people, and is positioned on the east side of the Ankara Metropolitan area.[85] Mamak, as one of the most famous and largest *gecekondu* districts in the periphery of Ankara, the heart of the Westernization project, had long been a symbolic space of "traditional" life styles, radical political resistance and poverty. As such, it has challenged the modernist image of the city during the fifty years of its *gecekondu* history. This constitutes one of the main reasons why it is identified more readily as an example of the *gecekondu* phenomenon in public perception than the other *gecekondu* districts of Ankara city. The region has been considered as a typical *gecekondu* district by the fact that the growth of the region exemplifies the peripheral nature of *gecekondu* expansion, the most commonly encountered pattern in Turkey. *Gecekondu* districts that have grown closer to the city center may not have witnessed all the processes of the growth and production of a typical peripheral *gecekondu* district and its capture of certain autonomy and spontaneity. On the other hand, Mamak *gecekondu* region, as a peculiar case, constitutes one of the rare examples to

85 See Mamak Belediyesi (2001), *1999-2000 Yılı Faaliyet Raporu.*

the unintended growth of a *gecekondu* district as a challenge from within to the radically modernist production of urban space. As a by-product of the highly modernist intervention in space in the service of producing a modern western city, the growth of the Mamak *gecekondu* region has witnessed, at the same time, the modernization history of Turkey. Locating at the gateway of many central Anatolian cities and towns to Ankara, the Mamak region started to absorb migration beginning in the early 1950s and this history will be discussed at length in the third chapter.

There are other diverse reasons behind the selection of the specific neighborhood as suitable for research interests. First, the locality hosts a large market place that is composed of nearly 150 shops and workshops. The market had long kept its status as the only market in a vast area and attracts customers from the majority of the neighborhoods in Mamak. The unanticipated presence of the market place comprised of a striking co-existence of many shops, selling national and international brands, alongside the very small shops of traditional craftsmen or those that sell goods particular to the needs of squatter life in a classical neighborhood of typical *gecekondu* houses, inevitably attracts the interest of a researcher who wants to conduct a study on spatial history. The market place is situated along a river valley on two sides of the main road that divides Boğaziçi and Fahri Korutürk neighborhoods. The spatial history of the market place has witnessed a transformation from a river that played a predominant role in the social spatialization of the neighborhood to a crowded market place that has provided a focal point of social interaction for years. The market place and the social dynamics that affected its development constitute the core of social spatialization in the locality. An understanding of these dynamics will hopefully highlight the interconnected nature of these peculiar or spontaneous developments in the locality and the context of urbanization in the city in general.

The second reason behind the selection of this particular locality is the socially heterogeneous quality of this setting, mainly in terms of hometown and sectarian differences. As will be supported by the empirical findings, these differences in social terms play a crucial role in the spontaneous social spatialization in this locality. In a qualitative type of research like this one, it seems crucial from the beginning to discover whether differences of gender, ethnicity or class also matter for those studied.[86] In this respect, particularly in the first phases of the research, I have paid close attention to any occasion where people explicitly talk about or describe their relationships in the

---

86 See Emerson, R., Fretz, R. and Shaw, L. (1995) *Writing Ethnographic Fieldnotes*, Chicago: The University of Chicago Press, p.135.

history of social spatialization, with reference to *hemşehri*, sectarian or ethnic differences. I have encountered, in most cases, a clear awareness and perception about their and "other" *hemşehri* and sectarian identities. In this respect, *hemşehri* and sectarian identities seem to explicitly come to the forefront of the narratives in most cases.

The third reason behind the selection of this particular setting is the 50 year history of the neighborhood that has witnessed the three periods of *gecekondu* in Turkey. The emergence and development of a classical *gecekondu* setting from the 1950s to present is quite dissimilar to those newly emerging illegal settlements of the 1980s that are usually defined as "*gecekondu*" but which show important variations from classical *gecekondu* districts. Most of the *gecekondu* houses in Mamak have been legalized because of their fifty-year history, by means of *gecekondu* amnesty laws. However, still in the designated research setting, and as a consequence of the inconvenient geographical landscape for construction of apartments and other social reasons[87], thousands of typical one-story *gecekondu* houses dominate the scene. Newly emerging illegal settlements in İstanbul, İzmir, Diyarbakır and Mersin have constituted a research interest for a number of studies in recent years. However, in order to make an alternative interpretation of official *gecekondu* history by posing challenges to it, there seems to be a need to choose a locality that has a long history and carries the traces of a classical *gecekondu* setting.

The *hemşehri* and sectarian identities are taken as important variables in this research constituting the main axis of competition and cooperation in the process of social spatialization. For that reason it is crucial to make explicit the common attributes shared by people from the same town of origin that symbolize certain political and sectarian affiliations. The choice of migrants to live close to their *hemşehris* makes the social heterogeneity of the region spatially detectable. This squatter region similar with other *gecekondu* settlements of Ankara inhabits migrants mainly from the villages and towns of central Anatolian provinces, namely Kırşehir, Kırıkkale, Çankırı, Çorum, Yozgat, Sivas, and the province of Ankara itself. There is also a large community of migrants from Tunceli who had settled in the region as a consequence of in-city migration in the early 1960s. At this point, it is worth mentioning that some provinces are categorized in a particular way depending on the provinces' identification with certain political ideologies

87 The resistance of local actors and other socio-spatial factors may constitute a barrier against the application of construction plans properly in spite of the legalization of *gecekondu* lands and houses in these neighborhoods. This issue will be discussed in detail in chapter six.

(leftist / rightist) and / or the dense presence of sectarian populations (*Sünni / Alevi*). Likewise migrants categorize each other in accordance with their hometowns.

The migrants from Kırşehir are the most densely represented migrant community in the region. The migrants from this community are all *Sünni* in sectarian origin and they are also known for their support to Islamist political parties starting from the early 1990s. In their everyday lives, they exhibit a conservative and religious lifestyle. Besides that, as will be mentioned in detail in the empirical chapters, due to socio-spatial factors, they are one of the most influential groups in defining the social space of the neighborhood. In spite of the fact that most of them have a center right approach in politics, they took no active part in the political struggles of the 1970s on side of ultra nationalist militant groups.

Migrants from Kırıkkale are the second most densely represented group in the region. They too are mostly *Sünni* in sectarian origin; however, there are also migrants from some *Alevi* villages of Kırıkkale in the region. Kırıkkale is still generally identified with nationalism in migrants' perception.

Third largest migrant groups are from Çankırı, Çorum, Sivas and Yozgat. These provinces are represented in the region by their heterogeneous sectarian structure. There are migrants both from the *Alevi* and *Sünni* villages of these provinces. As will be made explicit in the empirical chapters, the migrants of these provinces, regardless of their political inclinations, had suffered from the political struggles of the late 1970s relatively more in comparison to other *hemşehri* communities, which are homogenously represented in sectarian terms. *Sünni* migrants of Yozgat, both in their own or other migrants' narratives are identified mainly with ultra-nationalist and extreme rightist politics as with the *Sünni* migrants of Çankırı. Sivas is mainly identified with leftist politics and *Alevism* due to the dense presence of *Alevi* villages within its boundaries. The *Alevi* migrants of Sivas and Çorum seem to constitute communities that had either joined or given support to the leftist militant groups of the 1970s. Particularly, the *Alevis* of Sivas, due to their extensive support to leftist political struggle, are identified as the "other" of the neighborhood. This perception is not only accepted by the conservative *Sünni* communities but also by "acculturated"[88] *Alevi* communities of heterogeneously represented central Anatolian provinces.

The last significant community analyzed in this study is composed of migrants from Tunceli. These are distinguished from other migrants analyzed

---

88 Some *Alevis* of Çorum, Yozgat and Sivas due to the social pressure of their *Sünni hemşehris* go to mosque and fast during the Ramadan, which are not *Alevi* ways of praying.

in this study because of their Kurdish identity. In addition, in terms of their sectarian affiliation they are *Alevis* and they seem to be perceived as the most distant "other" by *Sünni* migrants. According to personal observation in the region, the social relations between migrants of Tunceli with other migrants retain the most distant quality within the general context of the socio-spatial composition of the region. They also seem to be physically the most isolated community as they had established themselves on the higher ground, topographically the most difficult place to settle in the region.

As it is clear from the above description of migrant features in the region, there is a high level of heterogeneity. This heterogeneity, within the context of the third research interest of the study mentioned above will hopefully provide a context to understanding the nature of different encounters among the members of these communities at some critical moments of social spatialization in the locality.

1.2.2 Methodological Framework

As discussed within the framework of conceptual and theoretical structure of the book, the main objective of the research is to trace the course of social spatialization in the locality. This will be achieved in the main through migrants' narratives about individual and neighborhood histories. However, it is crucial to note that this will neither be a detailed local history nor a report of all aspects of social spatialization and, in that sense, the historical account of focal points of social interaction in the neighborhood. Rather, it will be an endeavor to signify some moments of social spatialization that seems worth mentioning by the respondents as part of their individual history in the course of social spatialization. In this respect, these moments that may constitute examples of the research objectives of the thesis and in that respect some spaces, be they the components of market place, places of worship, village associations, schools or the critical places related to transportation would be brought into discussion to the extent they have played a role in the stories of the migrants. Rather than making a sample of these particular places in the locality and tracing all instances of social spatialization in them from the beginning, this will be a research of social spatialization through the narratives of migrants. Therefore some particular places were given greater space in comparison to others due to frequency of citation in the stories.

Individual narratives will be employed on two different levels in analytical terms. At the first level, the peculiarity of individual stories or the different perception or memory of the same incidents by each respondent is evaluated as crucial in understanding the nature of encounters between different groups in the course of social spatialization. This also gives clues to the dynamics of spontaneity embedded in this process. The respondents, in some cases, may manipulate history of the neighborhood and their place in it according to their own agendas. Different versions of the same incident may be given by individual respondents according to personal histories of migration, class, gender, religious affiliation, age and so on.[89] In the context of the book, these differences in interpretations will be analyzed to evaluate the way respondents situate themselves depending on their *hemşehri* and sectarian identities in the social spatialization of the locality.

At the second level, the totality of narratives in combination with my observations and interpretations constitute the building blocks of a personal interpretation of the actual course of social spatialization in the neighborhood. As mentioned above, researcher may encounter different interpretations of the same incidents. "These different versions might be grounded in some of the details not present in the other, to order actions in slightly different ways and to offer different interpretations..."[90] In this respect, the totality of the narratives may substantiate the author's construction of a story of social spatialization.

De Certeau states that one must locate a scientific model to explain the relation between narration and tactics (spatial practice), in which the theory of practices takes precisely the form of a method of narrating them.[91] In that respect, rather than searching for a secret or reality that exist behind all stories, for De Certeau, tales and stories should already be considered as practices. Ewick and Silbey[92], in their research mentioned above, accept a similar position with De Certeau in which case, the stories speak for themselves. They search for the examples of how respondents manipulate the structural means and system of the strategical realm to develop resistant acts. The authors find certain instances in the stories as constituting examples to the way respondents deal with law. In this book too, the significance and meanings respondents attribute to certain moments of social spatialization are allowed to determine the course of story. In that respect, rather than attributing an important place to certain components of public space from the beginning, the signified spaces that played a role in the stories of migrants were given predominance as mentioned above. However, beyond that, to appreciate the need for a second level of analysis that contains the

---

89 See Ley, David and Mountz, Alison (2001), "Interpretation, Representation, Positionality: Issues in Field Research in Human Geography" in Limb, Melanie and Dwyer, Claire (eds.), *Qualitative Methodologies for Geographers: Issues and Debates*, London: Arnold, p. 240.

90 Emerson, et al, (1995:117).

91 De Certeau (1984:80).

92 Ewick and Silbey (2003:1345).

interpretations and the conclusions of the researcher regarding the conceptual and theoretical framework s/he is dealing with seem to be necessary within the context of this research. This seems to be an important methodological concern shared by most researchers. Indigenous categories can only provide a starting point[93]. Over and above merely identifying these members's categories there is a need to specify the conditions under which people actually invoke and apply such terms in interaction with others. Shields shares a similar position in his spatial research, mentioning that pure hermeneutic would be inadequate if one is striving for a more broadly based sociological explanation.[94] In this respect, Giddens' double hermeneutic, which involves both the entering and grasping the frame of meaning adopted by the actors themselves in the production of social life, and the reconstitution of these within new frames of meaning according to the analyst's technical, conceptual schemas seems to be relevant with regard to the objectives of the study.[95] Insiders may ascribe to particular discourses regarding places as a mark of their insider status; however, there is a need for the researcher to understand the "real" dynamics underlying such discourses. As mentioned above, these dynamics can only be traced through the interpretation of the totality of the social spatialization narratives, as will be attempted in the book; sustaining balance between these two levels of analysis is a somewhat difficult task. An overemphasis of the scholarly framework would distort the narratives of the insiders; however, to simply present member's categories exclusively would produce texts devoid of relevance and interest to scholarly audiences.[96]

Considering these methodological difficulties, it is crucial to provide a thematic analytical spatial story of the research setting where key patterns, processes and regularities of social spatialization in the locality would be searched mainly in relation with the first research objective of the thesis mentioned in the section of research problem above. It is also important to tackle certain critical instances and moments in the course of social spatialization that may constitute examples relevant mainly to the second and third research interests defined above.

Besides considering "space" as an object of study in tracing some moments of social spatialization, the "spatial approach" is accepted as a methodological and analytical tool in the study. This approach, both as a theoretical and methodological tool, seems to suit reasonably well to the study of a *gecekondu* locality due to the particular qualities of *gecekondu*

---

93 See Emerson et al. (1995:139).

94 Shields (1991:26).

95 Ibid., p.20.

96 Emerson, et al. (1995:169, 170).

neighborhoods. The innate nature of the formation of *gecekondu* settlements necessarily incorporates a certain appropriation of and / or competition over space. The embedded spontaneity in these settlements is mainly related to this competition, as the findings will elaborate on. Spontaneity in the locality is also related to migrants' active agency in the process of social spatialization, in other words, the migrants simultaneously take the roles of planners, constructors and inhabitants of their space for a certain period. In this respect, the competition over and production of space constitute the primary concerns for the migrants in their everyday lives.

1.2.3 Research Methods

By applying qualitative research methods such as long semi-structured interviews, informal talks and observation as in the form of participating in the respondents' daily lives, the oral spatial history of the neighborhood and the way individuals situate themselves in this history were elaborated in the course of the research. The priority that is given to the individuals' narratives as mentioned above makes qualitative research techniques appropriate and relevant for the purposes of this study. The research started in the summer of 2002 and, with certain intervals, lasted until the summer of 2005. I paid daily visits to the region and endevoured to spend as much time as possible with respondents. In most cases, I tried to arrange more than one meeting with each respondent because in many cases, the things people say may not coincide with the things they really believe and certain amount of time is needed to form a trust relation as an "outsider" researcher. In this respect, the participation in the daily life of the respondents or to have informal conversations with them seems to be vital for the research porposes. In order to get "member's meanings", rather than asking directly about their ideas on certain issues, it can be more beneficial to observe and record naturally occurring talk.[97] Since some of the data concerns politically and religiously sensitive issues, informal talks and participatory observation have been more useful than formal interviews for the purposes of this research.

The "outsider" researcher role can also be acknowledged as an advantage for the fact that it sustains a certain comfort for the studied to express themselves in a more explicit manner on the other hand. In some cases, as Robina Mohammad[98] mentions, the positioning of the researcher

97 Emerson, et al., (1995:140).

98 Mohammad, Robina (2001), "'Insiders' and/or 'Outsiders': Positionality, Theory and Praxis" in Limb, Melanie and Dwyer, Claire (eds.), *Qualitative Methodologies for Geographers: Issues and Debates*, London: Arnold, p.109.

as an "insider" may also have an effect of making the researcher appear too close for comfort, making people wary of sharing information. Due to the anxiety on part of the respondents in making their past experiences in the neighborhood and ideas about certain issues explicit particularly with regard to their relations with "other" groups, the use of tape recorder was largely avoided, and informal talk was used to make respondents feel more comfortable.

One of the main objectives of the interviews was to learn about the life histories of the individuals in connection with the spatial history in the region.[99] Therefore, as will be mentioned in the section on sampling issues, older/more mature people who have histories in the locality had been given priority in selection for interviews or conversations. I have, mostly, conducted semi-structured interviews with one respondent at a time. However, I have not insisted on conducting individual interviews if there were a couple or a group of men or women who wished to express their experiences or opinions as a group. Particularly, in attempts to reach historical accounts, the conversations and debates within the group led to the surfacing of many memories, otherwise may remain tacit. In a similar research conducted by local history groups[100], the importance of group study in local history research is mentioned. As stated by Crang, remembering the past is a social undertaking, establishing and sharing senses of the past to produce "social memory". As the naturally occurring informal talks and disputes between the individuals might make explicit the real incidents and ideas implicit in the historical accounts of social spatialization, the intention behind the application of semi-structured interviews is to avoid interrupting the natural flow of the conversations. In order to manage the information with reference to the incidents and memories, which seem to be emphasized more by the respondents, I have restructured the interviews in the course of the dialogue.

Considering the heterogeneity of the region in sectarian terms, it is important not to form closer relations with people from any one of these sectarian groups than with the others. All stages of research have been facilitated by being completely explicit and honest about my identity as a researcher. Besides these, official maps of the neighborhood and the municipal documents related to the planning of the designated settlement constitute another source of data.

---

99 See Apendix A and B.

100 See Crang, Mike (2001), "Field Work: Making Sense of Group Interviews" in Limb, Melaine and Dwyer, Claire (eds.), *Qualitative Methodologies for Geographers: Issues and Debates*, London: Arnold, p.217.

1.2.4 *Sampling, Data Analysis and Some Weaknesses of the Research*

There are two main groups of respondents selected as appropriate to the objectives of the research. The first group comprises the migrants who can be defined as either the active or more powerful agents who have the ability to act near-strategically as defined in the conceptual framework. This group is mainly constituted by shop owners, who seem to be the most effective group and behave as the main power holders both in conjunction with their economic assets and networks in local politics. As a group, shop owners are also critical for obtaining the information on the history of social spatialization in the market that constitutes the core of the research due to the importance and centrality of the place as the focal point of social interaction. In this respect, shop owners constitute the primary group of population in accordance with the objectives of the research. In order to obtain a representative sample by considering the different functions shops perform, the country of origin and sectarian affiliation of the owners and the years spent in the neighborhood, the purposeful sampling[101] method is used. Since the research mainly traces spatial history, the shops and workshops that have a relatively longer history are given priority. 15 in depth interviews were conducted with shop owners. Five shop owners from Kırşehir were taken in the sample reflecting the dense representation of owners from this town. In addition, two shop owners from Yozgat, three from Ankara, two from Çorum, two from Çankırı and one from Kayseri were interviewed. Besides the interviews, I spent a certain amount of time in each shop to witness the daily routine and the relationship between customers and shop owners. During these visits, informal talks with the shop owners provided crucial knowledge in understanding the dynamics of social spatialization.

The ordinary inhabitants of the region constitute the second population of informants. 33 in-depth interviews were conducted within this group. Similar to the concerns as in the first group of population, the period of

101 "This is a strategy in which particular settings, persons or events are selected deliberately in order to provide important information that can't be gotten as well from other choices." Maxwell, Joseph (1996) *Qualitative Research Design*, London: Sage Publications, p.70. As defined by Maxwell, there are at least four possible goals for purposeful sampling. Firstly, the representativeness or typicality of the settings, individuals, or activities can be selected. Secondly, the heterogeneity of the population can be sustained. The third possible goal is to examine the cases that are essential for the theoretical and conceptual framework employed by the researcher. Fourthly, it can be crucial for enabling comparison, exemplifying differences between settings and individuals. See Maxwell (1996:71,72).

stay in the locality, hometown and sectarian affiliations constitute the main concern during the sampling process. In order to select the second group of respondents, a combination of random and purposive sampling method was used. Since the inhabitants from the same hometowns tend to settle close to each other, it is possible to clearly demarcate spatial concentration of migrants from the same *hemşehri* groups. Information was obtained from potentially suitable families in each *hemşehri* region, drawn from a random sample. In a sense, these randomly selected houses constitute keys to other families of the same hometown who have considerably longer history in the region and are from certain sectarian identity. In each region a number of families were selected using different randomly selected key informants, in order to avoid key informant bias. Qualitative researchers, as some cases show, may rely on a small number of informants either in the process of advising on how to take a purposeful sample or in the course of gathering the major part of their data.[102] Particularly, in *gecekondu* research, the key informant bias may imprison the researcher in a closed and small *hemşehri,* kinship or sectarian group; the main concern while choosing families in each *hemşehri* locality is, therefore, to reach houses those are spatially distant from each other. The main objective behind such a concern is to have a range of social spatialization in the broadest sense. A number of families were selected who were located near families of an opposing sect, in order to constitute examples to the processes and meanings of what it means to live in close contact with the opposing sect in the course of social spatialization.

By employing such a method, 13 *Alevi* and 20 *Sünni* respondents were interviewed. 16 of the interviewees are women while 17 are men. There are seven migrants from Kırşehir, six from Yozgat, five from Kırıkkale, four from Çorum, four from Tunceli, three from Sivas one from Kastamonu, and two from villages of Ankara present in the sample. Almost all of the respondents are above the age 40, and had spent considerable time in the neighborhood. In the composition of the sample interviewees, older people had been particularly sought. In this respect, migrants interviewed are the early arrivals to the region rather than the later arrivals and tenants. Therefore from the beginning there seems to be a bias in the research since almost all of the respondents are relatively powerful groups as the early comers who accumulate wealth, experience and social and political network in the city. A number of interviews with key people who have played important roles in the course of social spatialization had also been conducted.

Hometown and sectarian differences of *Alevi* and *Sünni* are taken as important variables that affect the relationship between groups in the

---

102 Maxwell (1996:73).

course of social spatialization, which have taken on different meanings and implications depending on the context of social spatialization. In this research, group identity or the definitions of "us" and "them" in the narratives of the respondents even in the beginning of the research were considered as context dependent and showing variations depending on spatio-temporal factors. Empirical data gathered in the later stages of the research supports a theoretical acceptance of the constructivist view of identity, which has been proposed, from the beginning of the research. In such a view, ethnic boundaries are the outcome of social interaction between groups. "The criteria that are used to categorize "Us" and "Them" are not objective differences, but features that the actors themselves regard as significant. Ethnic boundaries in this view are not given, but constructed, manipulated, subject to change and situational."[103] Even the simple definition of identity seeks this kind of reciprocity and construction rather than a consequence of given reality. Social identity includes our understanding of who we are as a group and our definition of the other group, reciprocally; and this is also true for the other group's definition of their identity.[104] However, it is important to mention that these definitions by their nature depend on the context of the social interaction. In this respect, identity can only be taken as a process; as being, or becoming. There needs to be a certain remembering and sharing in a group for the sake of the continuation of such collective identities. However, these rememberings and sharings may have shifting meanings over time; sometimes, one definition of "us" or "them" for a group may be contested by other definitions substantiated from another identity source for the same group. In this respect, collective identities are generated in transaction and interaction and are, at least potentially, flexible, situational and negotiable.[105]

The identity issue and hometown and sectarian commonalities as the most often mentioned identity references in the narratives are regarded mainly in such a constructivist manner throughout the book. It is also important to note that the basic objective of the study is not, in fact, to deal with the definitions or interpretations of identities. Rather I take identity definitions as variables or/and reasons behind the situational positioning of groups depending on the spatio-temporal context of the locality. Therefore, no detailed references to the theoretical literature on identity and no closer interpretation of and narrative analysis related to identity definitions will be made throughout the book.

---

103 See Erdentuğ, Aygen and Colombijn, Freek (2002), *Urban Ethnic Encounters: The Spatial Consequences*, London& New York: Routledge, p.10.

104 See Jenkins, Richard (1996), *Social Identity*, London& New York: Routledge, p.5.

105 See Jenkins (1996:102).

I initiated data analysis with a thorough reading of all the field notes and defined certain categories and themes through this reading. As mentioned by Maxwell[106], the key feature of most qualitative study is initial and continuing reliance on the data gathered in the process of defining codes for analysis in an inductive fashion. This inductive approach seems to be the dominant form in qualitative type of study in congruence with its nature since it is mainly grounded in the data. In research having an important historical component as this one has the initial categorization of the field-notes is greatly facilitated by putting the data into temporal units before a more specific and thematic coding. This also contributes to coming up with a thematic story at the last stage of the writing process. After the first and second revision of the field-notes, at the stage of initial thematic coding of the data, there emerged a need also to utilize existing literature and theory on which the research interest had been built. A coding process necessitates mainly a categorization of data, an overall review of existing theory in interaction with research data, and the consideration of conceptual structure of the people studied using emic categories. As Crang[107] mentions the quantification of qualitative data such as content analysis at times be a useful resource for tracing repeated memories, emotions, processes and incidents in local historical research. In the writing process, sometimes a need to reconsider the codes emerges. In turn, the data analysis also necessitates a reconsideration of the conceptual framework in an iterative fashion.

The most important difficulty in the field emerges from the perception of certain issues and certain moments as being rather sensitive by the migrants. These sensitive memories and issues particularly related to the late 1970s, when the struggle between leftist and rightist militant groups had radically politicized the *gecekondus*, make the process of gathering data rather difficult in some cases. Especially, related to this particular period, respondents are reluctant to share their experiences. They are also unwilling to make explicit their genuine beliefs and ideas about other sectarian and *hemşehri* groups and their tactical acts including informal moments in the course of social spatialization. In order not to take the risk of losing many instances of social spatialization by disturbing the natural flow of the conversation, I avoid using tape recorders in most cases. Rather, I take instant notes in the course of the interviews and prepare field-notes about my impressions of conversations and also about the attitudes and emotions of the respondents for each visit to the region. Therefore, I wrote up my field-notes of conversations as soon as possible, sometimes even on the bus home.

---

106 Maxwell (1996:79).

107 Crang (2001:227).

Second important difficulty is related to gathering information about religious spaces and practices particularly from the *Alevi* respondents. The *Alevis'* place for prayer *Cemevi*[108] is still not considered as a legally recognized place for prayer due to the domination of *Sünni* interpretation of Islam in state policies about religion. There is only one *Cemevi* in the region and that is not recognized officially, but informally functions as a cultural foundation for *Alevi* people however, they conduct religious gatherings in it. My participation in such gatherings and recording of detailed information about the construction of *Cemevi* has been a difficult process. Moreover, behaving insistantly in order to gather knowledge might have caused suspicion within both *Alevi* and *Sünni* communities. This might also have restricted the possibility of gathering knowledge from people about other, less sensitive issues.

The formal places of prayer for *Sünni* population are the two mosques in the settlement. Since the mosque is mainly used by men, the informal religious gatherings at home constitute an important part of the religious life for *Sünni* women. However, these informal gatherings at home have also been under close surveillance of security forces. In this respect, I could only participate in two religious gatherings at home. Since it is difficult to define public and private realms in *gecekondu* settlements, such gatherings are also considered as public issues therefore considered as important parts of recent social spatialization in the neighborhood. However, *Sünni* women are also reluctant in sharing information about these gatherings. Village associations of *Sünni* migrants, which are usually used as coffeehouses by men due to their status as masculine spaces, meant that it was very difficult for a female researcher to sustain long hours of participation in the daily routines there. In spite of all these difficulties, not being over-insistent on attaining certain knowledge and entering in certain spaces had helped me to build relationship of trust and to gather in-depth knowledge about other aspects of social spatialization, which in turn revealed, important clues about these sensitive issues in the course of the research.

---

108 The *Alevis* do not use *Cemevi* in the way orthodox *Sünnis* use mosque. Unlike a mosque or church, *Cemevi* is not a place where one conducts prayer in order to comfort her/ his conscience. In order one to go to *Cemevi*, at first, one need to be peaceful. *Cemevi* is a multivocal place. In most of the *Alevi* villages, *Cemevi* is used for purposes other than religious ceremonies in normal times. See Çamuroğlu (2000:83).

1.2.5 An Outline of the Study

The main contribution of this study is to reveal some instances of the interdependent nature of the formation of *gecekondu* settlements and planning of city space in general, employing a new framework to appreciate the reciprocity between tactical acts of *gecekondulu* and the agency of strategical realm. The *gecekondulu,* rather than being a marginal actor tactically, employs the structural means of the strategical realm for the sake of constructing informal *gecekondu* space. Any planning practice that wants to attain success should consider local knowledge in these neighborhoods, where spontaneity plays a crucial role in social spatialization. Spontaneous and contingent development in these settlements is very much an embedded part of local social dynamics between different groups in the locality. Therefore, it is to be hoped that, a sketch of social spatialization with a retrospective perspective in one of the oldest and most typical *gecekondu* settlements will shed light on these local dynamics, which have taken different forms over time. Local knowledge, and the ways recommended to attain it in this study could potentially produce relevant information of interest to those involved in innovation in urban planning and politics. This way of gathering local information may also contribute to a new understanding that shifts the analytical focus from a conventional understanding of urban governance to a more democratic one.[109] Democratic and more inclusive methods of local governance will be possible by piecing together local knowledge.

With such a conceptual and methodological agenda in mind, the second chapter is devoted to the critical historical sketch of how *gecekondu* neighborhoods have been represented in public discourse in parallel to the socio-spatial developments in them. In this process, the hesitant approach of politicians to the *gecekondu* issue will be problematized within an overview of *gecekondu* laws and politics. The third chapter includes a critical and detailed reading of the planning and construction of Ankara city and the emergence of spontaneous spaces in the periphery of the city as a paradox and impediment of this planning activity. The initial settlement years in the designated research setting will be elaborated in chapter four. The settlement decisions of migrants, construction of *gecekondu* houses and the development of the basic socio-spatial schema of research setting will be elaborated with reference to the dialectical interaction between migrants' tactical acts and natural factors as the most important spatial determinants of the time. Chapter five is devoted to an analysis of the political turmoil of the 1970s

109 Groth&Corjin (2005:504).

in the neighborhood. The effect of the blatant and harsh struggles between leftist and rightist groups will be portrayed by an examination of radically political social spatialization in the locality. The changing context of power relations due to tactical wars on the horizontal level will also be considered. In chapter six, the changing context of social spatialization due to neoliberal policies and identity politics beginning in the 1980s will be examined. It will be a showcase to explicate the transition from a "space of politics" to a "politics of space" in the course of social spatialization in the neighborhood. The spontaneity and contingency embedded in spatialization during that period will be extensively evaluated and the religiously dominated identity perceptions will be elaborated in relation to the socio-spatial context of the period.

CHAPTER 2 - Strategical Realm's Discourse Of *Gecekondu* Settlements

2.1 Introduction

The sixty years history of *gecekondu* settlements in Turkey is considered as the most important side effect of the rapid urbanization process, which has been itself embedded in the political and economic context of the country sharing similar and in some cases, common instances with many developing countries' histories. During the long history of *gecekondu* in Turkey, however, different interpretations of this factor have delineated various ways of representing these settlements. As briefly mentioned in chapter one, the system-related causes of *gecekondu* construction in Turkey have been underemphasized in most early academic research and also in public debate. This approach has attributed an overburdening responsibility to the migrants as the main aggravators of the mal-urbanization in Turkish cities. On the other extreme, the modernist public approach had attributed a totally effective agency to the actors who have the authority to represent space in the strategical realm. In so doing, the acts of representing, producing, tabulating, planning and imposing spaces by a collection of agents such as urban planners, politicians, state bureaucrats and scientists have been considered as adequate to effectively transform *gecekondu* settlements into modern districts over time. Such an approach neglects potential spontaneity embedded in the social spatialization of *gecekondu* space, which depends considerably on the capabilities of local agency.

Considering the weaknesses of these two opposing approaches, the socio-spatial transformations in *gecekondu* settlements will be dealt with by taking the changing political and economic context of the country into account. The simultaneous analysis of the legal context in dealing with *gecekondu* spaces and the changing representations of *gecekondu* space will constitute the core of this chapter. This objective will be accomplished by dividing *gecekondu* history in Turkey into three main phases showing certain correspondences with the phases of social spatialization as signified by the respondents of the research.

The first period covers the years 1940 to 1966. This period signifies the first encounters with *gecekondu* settlements. Though *gecekondu* settlements had become widespread through the end of this period and the concept of *gecekondu* had been used in public debates that criticize the policies regarding these settlements, the "*gecekondu*" were not formally acknowledged until 1966, the year when the first *gecekondu* settlement law was issued. The period covers Republican Regime's policies of constructing a modern life style through urban planning and also the transition to multiparty politics, which had altered the policy of urbanization in a dramatic manner.

The second period includes the years 1966 to 1980. This period covers the most dynamic peirod in the socio-spatial transformation and the rapid spread of *gecekondu* settlements in the periphery of cities. This period witnessed the ossification of *gecekondu* settlements in the physical, political, economic, social context of large cities in spite of the fact that they had been represented as marginal spaces in public and academic debates. Political and economic instability in the 1970s had led to radical struggles between extreme leftist and rightist militant groups in all towns and cities of the country. *Gecekondu* spaces have also been subject to this struggle, even more than the other places, due to the close encounters between *Sünni* groups claiming extreme rightist and ultra-nationalist ideologies and *Alevi* groups appropriating extreme leftist ideologies in the neighborhoods. As mentioned in the introduction chapter, these were the years in which the homogenous identity imposed on *gecekondu* society as the "rural other" was completely overturned as a consequence of the conflict-ridden encounters between different identity groups in *gecekondu* settlements.

Third period starts with 1980, the year of military intervention, carried out with the expressed aim of restoring democracy in the politically unstable context of the country. It paved the way for a new phase in Turkish history, which was to become dominated by the neoliberal policies of the first civil government. This neoliberal approach to economy and politics transformed the whole estate system, which also included *gecekondu* neighborhoods, in their context of transformation. The socio-spatial transformations in these settlements had hardened the content and intensity of exclusionary representations of *gecekondu* neighborhoods in public debate. The intense proliferation of illegal settlements in the periphery of large cities after the 1980s, which greatly contrasts with the socio-spatial emergence and qualities of classical *gecekondu* settlements, has complicated the issue of the analysis of representation of low-income settlements.

2.2 1940-1966: The Period of First Encounters

2.2.1 Rural-to-Urban Migration and the Initial Emergence of Gecekondu Settlements

The period between 1940 and 1966 witnessed the implementation of important economic and structural reforms with the aim of integrating Turkey into the world economy. The single party regime that had lasted until the late 1940s sought statist economic and cultural policies, which aimed to construct a national industry based on the country's own resources, and implemented a civilized life style modeled on the West. The Republican People's Party's (*CHP*) statist, centralist and protectionist economic policies of import substitution and intense control of agricultural production including the forced collection of farm produce alienated especially the landlords and peasants. This alienation had prepared the context for a warm welcome and popular support for Democrat Party (*DP*), the victor of the first elections of multi-party period held in 1950.[110] As an alternative to the statist policies of *CHP*, *DP* seemed to integrate a more liberal approach to economy and society. The first seeds of the process of integration with west through Truman Doctrine and Marshall Plan had already been seen during the single party era. *DP* continued on the integration, which constituted the core of its party program and the party played a major role in this integration process. However, in time, it was accused of carrying similar policies with *CHP* in alienating masses with its repressive, populist and hesitant policies. During the *DP* rule, the control of economy was to a large extent left to western powers; economy was dependent on the export of raw materials and food to Europe [the demand for food was high during post war years in Europe, but declined in succeeding years], and on highly mechanized but ineffective and unproductive agriculture.

The lack of agricultural productiveness stemmed from the continuing dominance of landlords at the local level and unstable populist economic policies at the national level. With the aim of overcoming landlord domination in the countryside, *CHP* put a land reform bill into effect in 1945. This bill attempted a dramatic structural transformation in traditional means and relations of agricultural production; it aimed at reducing the negative effects of local and parochial loyalties mainly among rural population and breaking the political hold of the landlords in rural regions by transforming Turkey into a republic of independent peasant proprietors.[111] However, due to the populist bias of *DP*, the conditions of this bill could not be properly implemented. The bill that was brought into being with the object of the distribution of privately and state owned land to peasants and the improvement in their conditions, actually worked against them as a result of the political pressure and the power of landlords.[112] Another factor that further complicated and hardened the conditions of peasants and turned most of them into landless

110 See Ahmad, Feroz (1996), *The Making of Modern Turkey*, London: Routledge, p.102.

111 See Ahmad (1993:103).

112 "Between 1947 and 1962 only about 1.8 million hectares were distributed to 360,000 families with only 8,600 being taken from privately owned land. The peasants again lost. The state owned lands, which were distributed, had been essential to sustain the landless or near-landless peasants who had used them for communal grazing. These people were reduced to the status of farm laborers, or they migrated to cities in search for work." Ibid., p.116.

farm laborers was the intense mechanization of agriculture, with the introduction of 42,000 tractors into the Turkish agriculture by the Marshall Plan. This further worsened the conditions of peasants vis-a-vis landlords by making them dependent for technology on landlords in addition to land. This change in the social balance in the countryside played an important role in the rural-to-urban migration. Another important factor contributing to the scale of rural-to-urban migration was the expansion of transportation facilities with the construction of road networks enabled by US aid during *DP* era. This development increased the mobility of peasants and enabled them to reach nearby towns and cities.

Two different spatial models of urbanization emerged out of these developments. First model of urbanization emerged as a consequence of the increasing visits of villagers to towns nearby. These encounters had enlivened the economic life in terms of trade based on agricultural goods and led to the gradual settlement of the villagers in these towns temporarily or permanently. This transition caused an increase in population in previously small towns, bringing in them city status. Erol Tümertekin[113] considers Turkish urbanization, preeminently, as a consequence of this transformation. This trend brought a relatively balanced distribution of cities and population in the 1950s. The urbanization of the country, particularly in the late 1940s and early 1950s, was the result of the growth of the small towns, which acquired city status. These small cities had been the first stop for the migrants on their way to larger cities in the west, İstanbul, Ankara and İzmir. The second spatial model of urbanization owes to intense rural-to-urban or urban-to-urban migration beginning by the late 1950s and continued through the 1960s. This led to the rapid growth of the largest industrial cities, paving the way for uneven regional development.

The early phase of the city life for the migrants was a transitory one. Usually the male members of the households migrated to the city while leaving the women and children back in the village. There were mainly two patterns of migration in the 1950s. First one is the "migration from village to a smaller town then to a city, namely indirect migration; and the second one is the "migration from village or small towns directly to a city."[114] Within the context of the indirect migration, there appears an extended space for the migrants to accumulate knowledge and experience regarding social

113 See Tümertekin, Erol (1973), *Türkiye'de Şehirleşme ve Şehirsel Fonksiyonlar* (Urbanization and Urban Functions in Turkey), İstanbul Üniversitesi Edebiyat Fakültesi Coğrafya Enstitüsü Yayınları, p.120.

114 See Karpat, Kemal (1976), *The Gecekondu: Rural Migration and Urbanization*, Cambridge: Cambridge University Press, p.26.

networks and connections in the large cities during their stay in the towns. Indirect migration also brought about the first model of urbanization mentioned above, contributing to the transformation of small towns into cities.

In both of these migratory trends, migrants protected strong links with their villages and if they had a land in the village, it is possible to talk about a double income for the families. With the increasing familiarity of migrants with city life, the migration took a more permanent character. The migration process that was directed mainly by the "push factors"[115] started to take more permanent form, directed by the combination of push and pull factors. This transformation led to a chain type migration in which the forerunner migrants in the city share their knowledge and material sources with their *hemşehri* -fellow countrymen- in order to help them to settle as neighbors in the city. As will be discussed in chapter four, the migration dynamics that contributed to the emergence of the designated research setting do not fit exactly with these defined migratory trends. Interestingly, rather than urban employment, the urban agriculture in the periphery of Ankara had led to a seasonal form of migration in the 1940s that constituted the basis of permanent migration of families in the late 1950s. As defined by İçduygu and Ünalan[116] there are multiple forms of internal migration in Turkey. These are interregional seasonal migration from rural to urban places; seasonal rural to rural migration; migration from urban to urban; migration from urban to rural and particularly after the late 1980s, the forced migration from Eastern parts of the country.[117] The form of early migration in the designated research setting seems to be seasonal rural-to-"rural periphery" of Ankara.

In sum, the reasons behind the rapid urbanization consist of agricultural mechanization and structural transformations following, such as the encouragement of cash crops production by state, the changing social relations in rural settlements working to peasants' disadvantage while making landlords relatively more powerful socially and economically;

---

115 Push factors can be defined as the insufficient living conditions in the rural context. Pull factors can be defined as the attractive opportunities of the urban environment. Pull factors were the facilities and services urban context provided for the migrants. These can be listed as the education and health facilities, the wide range of employment opportunities and entertainment facilities in cities.

116 İçduygu, Ahmet and Ünalan, T. (1997), "Türkiye'de İcgöç: Sorunsal Alanlari ve Araştırma Yöntemleri" in İçduygu, A. (ed.), *Türkiye'de İçgöç: Sorunsal Alanlari ve Araştırma Yöntemleri Konferansı* (Internal Migration in Turkey: Problem Areas and Research Methods), Türkiye Ekonomik ve Toplumsal Tarih Vakfı, p.41.

117 See Ibid., p.41.

better opportunities of health and other social service provision in cities; greater opportunities of mobility, due to both governmental policies that were permissive rather than oppressive towards migrants, and the construction of roads that made people more mobile.[118] What makes Turkish *gecekondu* formation a notable and rapid experience was the distinct quality of land ownership patterns in Turkey. There is much more state land than the privately owned land, in the periphery of Turkish cities, which can be considered as a consequence of the land system inherited from Ottoman Empire.[119] It is important to remember the fact that even in the early periods of rural-to-urban migration in Turkey, the *gecekondu* population in three largest cities of the country namely Ankara, İstanbul and İzmir constituted as much as 59%, 45%, and 33% respectively of total city populations.[120]

These developments led to the emergence of *gecekondu* settlements starting from the 1950s in the periphery of large cities. However, before that, even during the single party era, it was possible to encounter illegal shelters with very poor conditions in large cities. According to the journalistic interviews conducted in Altındağ neighborhood of Ankara in 1947, 40,000 people were estimated to live in cave-like shelters there, covered with tin roofs with small holes to get light in.[121] The emergence of these settlements was the consequence of the beginning of migration mainly as a result of push factors. In those years, *CHP* practiced tight state control of industrialization; therefore, for the migrants there had been no or little vacancies in the sector of heavy industry, which was rarely present in most cities. There was also no service sector employment during that period in the largest Turkish cities. Migrants, whether seasonal or permanent, had settled either in shelters described above, or rented rooms with desperate conditions in certain low-income neighborhoods in the center of the cities.

There were three low-income groups in Ankara living in illegal settlements or under difficult conditions. The first group was the low-income civil servants employed in Ankara's large bureaucratic sector. The second consisted of

---

118 See Danielson, Micheal and Keleş, Ruşen (1985), *The Politics of Rapid Urbanization*, London: Holmes and Meier, p.33. See also Kıray, Mübeccel (1998), Kentleşme Yazıları (Writings on Urbanization), İstanbul: Bağlam Yayınları

119 See Friedrich Ebert Vakfı (1996). *Türkiye'de Gecekondunun 50. Yılı* (The 50th Anniversary of *Gecekondu* in Turkey), İstanbul: (FES) Ozan Publications, p.5.

120 See the reports on *gecekondus* in Ankara, İstanbul and İzmir that were prepared by the Ministry of Reconstruction & Resettlement General Directorate of Housing in 1966.

121 Şenyapılı, Tansı (1998), "Cumhuriyet'in 75. Yılı, Gecekondunun 50. Yılı" (The 75th Anniversary of the Republic and the 50th Anniversary of *Gecekondu*) in *75 Yılda Değişen Kent ve Mimarlık*, Tarih Vakfı Yayınları, p.302.

rural migrants from the nearby towns and smaller cities of central Anatolia. The third group was the native Ankara dwellers or rural migrants resettled by state forces as a result of war, natural disaster or in the process of the construction planning of Ankara city.[122] These groups were the main actors of *gecekondu* construction in Ankara. As was the case in İstanbul, Ankara's *gecekondu* settlements grew intensely towards the end of 1950s.

Considering the emergence of illegal shelters in Ankara, İstanbul and İzmir as early as the late 1940s and early 1950s, in the period of innovative and determined attempts by Republicans to create a modernist urban space in physical and social terms, it is important to reemphasize the criticisms posed to Lefebvre's and De Certeau's stress of the power of strategical realm versus the tactical realm. As mentioned above, the modernization and industrialization policies of governments through a statist agenda during the Single Party era inevitably led to the transformation of the structure of agricultural production. In fact, these statist policies of modernization implemented by the *CHP* particularly towards the 1950s paradoxically also appealed to certain populist aspects and concerns, to avoid the alienation of influential social classes. These populist concerns continued to gain impetus with the *DP*'s ascendance to power.

Even in the early years of Republican Regime, the populist concerns of political actors had fragmented the modernist and homogenizing discourse and practice of the strategical realm. This fragmentation had taken a more intense form during the *DP* era, as a consequence of the increasing responsiveness of political actors to migrant masses in the city. The emergence of illegal shelters in big cities has various causes: rural-to-urban migration as a consequence of agricultural reforms; the insufficiency of the industrial sector to provide jobs for the migrants; and the speculative nature of the urban real estate system in big cities that had prevented the development of efficient social housing policies. The legal measures regarding urban problems particularly beginning by the *DP* period triggered the expansion of these illegal shelters. The initial fragmentation in the strategical realm was caused mainly by means of political actors' populist bias joined together with the strong and ardent modernist discourse on part of state bureaucrats and academics in relation to rural-to-urban migration.

2.2.2 Legal Strategies with Regard to Spontaneous Spaces: Un-Stating Gecekondu

DP came to power with the support of the masses and, in cities, the votes of *gecekondu* dwellers who had been alienated by the policies of *CHP* for many

122 Ibid., p.304.

years. Consequently, *DP* could not stay irresponsive to the demands of *gecekondu* settlers, a condition that led to the party adopting more populist policies. Therefore *DP* era is to be considered as the breaking point in *gecekondu* history, transforming the form of *gecekondu* settlements from randomly scattered houses or sheds as seen in the late 1940s, to more permanent and established neighborhoods in the 1950s.[123] In an environment where the old "modernist" city inhabitants, the media and governors all were in strong opposition to the emergence of *gecekondu* settlements, Şenyapılı points out that politicians tended to overlook the rapid growth of these settlements. She adds that this was not only related to the vote potential of these settlements, but also to the economic functions performed in large cities from the 1950s onwards.[124]

Before *DP* era, during the single party regime, the illegal settlements referred to in construction laws, but the political measures were unable to produce any significant effect. The state tried to prevent construction and social events taking place outside the framework of the planned urban space. The statist and regulatory policies had shown themselves in the form of legal arrangements that dealt with eradicating the preliminary symptoms rather than the causes of mal-urbanization during those years.[125] For example, between 1930 and 1950, five major laws dealing with the land and housing issues and aiming at preventing the illegal house construction were implemented.

The first group of laws aimed at the improvement of the real estate system by making some re-organizations in the ownership patterns of the state lands. In order to halt land speculation and to overhaul the real estate market, the laws provided for the transmission of the possession rights of these state owned lands to municipalities. In this way, the municipalities were obliged to buy the state lands on credit, develop it according to plans and resell it.

The initial law was complemented by two new laws in 1948. The first foresaw the transition of the ownership of some treasury lands to Ankara Municipality on credit that would be paid in installments over ten years. In so doing, the law enabled Ankara Municipality to allocate and transfer part of its land, under special circumstances, to people who wanted to build their own houses.[126] A complementary law was passed in the same year, encouraging

---

123 Şenyapılı (1998:309).

124 Ibid., p.310.

125 See Heper, Metin (1978), *Gecekondu Policy in Turkey*, İstanbul: Boğaziçi University Press, p.12.

126 Ibid., p.18.

house construction by municipalities, and extending the jurisdiction of the previous law to all the municipalities in Turkey.[127] The first successful example of social housing, the construction of Yenimahalle district was enabled by means of these two laws. The second group of regulations stipulated the demolition of houses built without construction permits. These laws were promulgated after the municipal law, which shows that these previous laws had been unable to prevent illegal housing construction.

The law legislated for the demolition of buildings without construction permits came in 1949. It directly targeted *gecekondu* houses, which seemed more visible and permanent to those people who believed that these settlements should not become permanent. Second law was established in 1953 and envisioned the provision of social housing to the ones who had built unhealthy houses without permits.[128]

However, due to the lack of resources and technical manpower, these laws could not be effectively implemented, in the same way that succeeding laws in the following five decades would not be realized, but with the addition of populist concerns. The ineffectiveness and inability of the strategical realm to handle these spontaneous spaces was not only related to the internal impediments and contradicting interests of the actors in this realm but also to the changing socio-political conditions of *gecekondu* settlements. The number of *gecekondu* houses had increased dramatically, particularly in the late 1950s and early 1960s. It became difficult for state officials to apply police measures in a context where migrant resistance was becoming more determined, with high degree of solidarity among *hemşehri* groups. This was the period in which *gecekondu* settlers had also become aware of their political bargaining power in demand for electricity, water and other infrastructural services for their neighborhoods. More and more political actors therefore began to adopt a populist approach towards *gecekondu* dwellers as their bargaining power increased.

The internal impediments of the strategical realm had been compensated for other factors such as the clash of interests of the actors in the strategical realm. At the local level the civil servants in charge of the implementation of laws in relation to illegal housing had a tendency to overlook some cases. This was an important phenomenon, which I myself encountered in the course of the research. There were many reasons for this. First, since some of the civil servants themselves were *gecekondu* dwellers, it was difficult for them to implement the demolition decisions. They did not harm their fellow *gecekondu* dwellers with who they shared the same fate. Second, for many

127 Ibid., p.18.

128 Heper (1978:20).

of the civil servants, the demolition process became a means of obtaining economic benefit by taking bribes from the migrants. Third, in some rare cases a strong resistance in *hemşehri* or neighbor groups met the officials, which greatly hindered the implementation of the demolition decisions. In some cases, the demolition decisions were implemented only partially, for instance, the partial rather than complete destruction of houses, which left buildings in tact and habitable.

During the *DP* era, three construction laws were promulgated, and these laws provided official amnesty for illegal low-income settlements which had previously emerged in the peripheries of the cities. These laws specifically avoided using the label of *gecekondu,* widespread in public use during those years. There was a common condition in almost all of these urban planning laws.[129] The amnesty was valid for the illegal houses that had been built before the law became effective. In spite of this condition, however, the laws by no means had any dissuasive effect. As a consequence of populist policies, migrants could easily make future estimations that governments would make new amnesty laws with almost every coming election. However, despite *DP*'s initial populist policies and their effective role in the increasing proliferation of *gecekondus* in the periphery of the cities, *DP*'s approach to urban planning took a "despotically modernist" turn particularly towards the end of their ruling period. As in the Single Party regime, *DP*'s aim was to erase traces of the past and poverty at this time. Accordingly, the party implemented formalistic policies to make Ankara and İstanbul resemble to modern European metropolises.[130]

As this section reveals, the agency of the strategical realm has not represented a uniform attitude and intention with regard to low-income settlements. This inconsistency has two basic causes: contradicting concerns of the political actors and state bureaucrats, and the changing social and political context over time. In this initial period, the legal measures that were taken in relation to urban problems and planning were mainly formalistic

---

129 "During the period between 1948 and 1966 three construction laws have been issued. In 1953 law number 6188 encouraging apartment constructions was put into effect and forgave illegally constructed houses that have been built until that time. Following that law number 7367 issued in 1959 which foresaw the distribution of state owned lands out of the municipal boundaries to poor people. In 1963, another law that guarantees service provision to houses, which were officially lacking municipal permission for residing was issued." See Ekinci, Oktay (1998), "Kaçak Yapılaşma ve Arazi Spekülasyonu" (Unlawful Construction and Land Speculation), *75 Yılda Değişen Kent ve Mimarlık*. İstanbul: Tarih Vakfı Yayınları, p.194.

130 See Şenol Cantek, Funda (2006), *Sanki Viran Ankara,* İstanbul: İletişim Yayınları, p.51.

in nature and contain a certain disregard of illegal shelters emerging at the center and periphery of the cities. This was the result of legal indifference combined with the strongly modernist approach on the part of state elites in their attempts to represent the modernist urban space by indicating the "other" of this space, namely *gecekondu* houses and *gecekondulus*.

2.2.3 *The Elitist Approaches to the Initial Formations of Gecekondu Settlements*

Concern was caused by the emergence of *gecekondu* in the periphery of cities, especially in Ankara, which was at the core of the Republican Regime's plan to impose a western lifestyle. There was a great deal of discomfort and prejudice regarding the rural, "backward" lifestyle there, as compared to the mainstream of the city. This concern was expressed both in public and academic circles in the 1950's and 60's. Such a strong reactionary representation of low-income settlements in the center and periphery of the cities, and a neglect of the system-related causes of their emergence had gone together with an optimistic belief that these settlements were merely transitory, and would eventually disappear. Parallel to the policies dealing with the problems of urbanization in a rather disingenous manner as mentioned previously, the public approach had foreseen possible solutions by eliminating the visible consequences of the problem, *gecekondu* settlements, rather than deal with the problem itself. In spite of its repeated failures, a formalist approach in urban planning and policies has been dominant from the 1940s until present. Between 1955 and 1960, particularly, wide boulevards were constructed in Ankara and İstanbul under the name of construction movement.[131] This can be considered as part of the hesitant approach of *DP*, squeezed between its early populist and later modernist policies. The elitist bias fed on the modernizing ideals of the Republican Regime, had taken a more concrete and conservative form during those years. It may be considered as a result of the unease on part of the "urbanite" sector of the society, who had to a high degree internalized the modernization principles and values regarding the increasing visibility of migrants in cities. As Önder Şenyapılı argues, the people who drew attention to the system related causes of migration and *gecekondu* construction had been held subject to legal investigation in the early periods of *gecekondu* construction within the framework of this elitist approach. The discomfort on part of the elites about the presence of *gecekondu* settlers in the city, first articulated in the late 1940s developed into a systematic prejudiced representation, which would continue to have effect

---

131 See Şenyapılı, Önder (1981) *Kentleşemeyen Ülke Kentlileşen Köylüler* (The Unurbanized Country, Urbanized Villagers), Ankara: ODTÜ Mimarlık Fakültesi Yayınları, p.8.

for years. Metin Toker, a journalist, in an article published on 28 September 1948, pointed out, in a somewhat forceful manner, the existence of *gecekondu* settlements side by side with İstanbul's most refined and beautiful middle class neighborhood Şişli.[132] One year later, the same journalist, referring to İstanbul city governor wrote: "Governor (Fahrettin Kerim) Gökay seems to protect *gecekondu* constructors; there is nothing to say. However, the *gecekondus* have their own laws, order and administration today, if [they] create their own police forces or military tomorrow, it will come as no surprise."[133] The academic and public representation in the following years defined *gecekondu* development as a threat. These were the initial years of *gecekondu* construction, characterized by a wishful, optimistic discourse on the transitory and reversible character of these settlements, and the migration process. At this time some proposals went as far as suggesting restricting access to large cities, or granting visa for entering İstanbul and Ankara.

The belief in the transitory and rural character of these settlements was expressed also in academic works. Research conducted by İbrahim Yasa in 1964 captured certain stereotypical representations of migrants with reference to their rural life styles in a prejudiced manner.[134] Migrant women working in middle class houses, such as domestic laborers were referred to as scrap dealers because of the fact that they wore the discarded garments of middle class families over their rural clothes, which was seen as extremely inelegant and threatening to the modernist perception of aesthetic.[135] According to the premises of the study, *gecekondu* families had to integrate with city life in order not to be seen as a threat to security in the urban environment. The inherent assumption was that *gecekondu* families who were in the middle of their process of transition caused social unease and insecurity in the city. The eating, dressing and entertainment habits of *gecekondu* family were described in a discourse of prejudice in order to mark the "rural other" of the modernization process. The strongly elitist and modernist ingredient in academic studies continued its existence in the following decades with an emphasis on the necessity of integration of these settlements into city space both in physical and social senses. Since the academic and public approach was dominated by this problem of integration, it neglected the dynamics behind the rapid growth of *gecekondu* settlements in the 1960s and 1970s.

---

132 See Şenyapılı, Tansı (1998:308).

133 Ibid., p.308.

134 See Yasa, İbrahim (1970), "*Gecekondu* Ailesi: Geçiş Halinde Bir Aile Tipolojisi," (*Gecekondu* Family: The Family Typology in Transition" *ASBF Dergisi*, 25(4), pp.9-18 and also (1973), "*Gecekondu* Ailesi Mozayiği," *Amne İdaresi Dergisi*, 6(2), pp.41-47.

135 Yasa (1970:11).

2.3 1966-1980: The Period of Settled *Gecekondu* Neighborhoods

2.3.1 Rapidly Spreading and Socio-spatially Transforming Gecekondu Settlements

The first formal acknowledgement of squatter settlements using the commonly-known name occurred with the "*gecekondu* law" in 1966. It seems to reflect the collapse of the belief in the transitory character of these settlements in the periphery of the cities on the discursive level. This change of approach had a strong basis in the facts. The percentage of *gecekondu* population in the total urban population increased rapidly during these years. In the year 1955, the share of *gecekondu* population in city population was 4.7%; in 1960 this percentage increased to 16.4% and in the following five-year period the percentage grew from 16.4% to 22.9%.[136] If we consider these percentages just in large cities like Ankara, for example, we see that, in 1955, the percentage of *gecekondu* population in the city's total population was 21.8%. This percentage shifted to 56% in 1960, 57.4% in 1966, 60.6% in1970, 64.9% in 1975, 68.4% in 1978 and 72.4 in 1980.[137] It is possible to argue that, during this period *gecekondu* settlements experienced an immense growth incomparable with other periods in *gecekondu* history. This period not only witnessed the growth and increasing density of these neighborhoods but also the maturation of *gecekondu* dwellers' experiences in the city environment.

Beginning from the national elections of 1969, in and through the national and municipal elections of 1973, 1975 and 1977, migrants played more effective roles with the population increases in these settlements. The effectiveness of right-wing politics in local government had been replaced by the domination of left-wing politics with the support of *gecekondu* dwellers.[138] Politicians could not remain irresponsive to the demands of *gecekondu* population for two reasons. First of all, the population of these settlements had shown serious increases during two above mentioned period. Particularly in local elections, *gecekondu* dwellers alone had the power to determine the fate of elections. Secondly, *gecekondu* population gained enough experience in the city environment during those years in order to use their political power tactically for bargaining purposes in attaining infrastructural services. This was the period signifying the beginning of reciprocity between the tactical and strategical realm that would gain certain intensity in the following decades, as will be discussed in depth in chapter six by referring to research data.

---

136 See Keleş, Ruşen (1993), *Kentleşme Politikası* (The Politics of Urbanization), Ankara: İmge, p.383.

137 Ibid., p.384.

138 Keleş (1993:389).

This phase in *gecekondu* history also witnessed the deconstruction of the homogenous identity imposed upon *gecekondu* dwellers as the "rural other" by the elitist discourse. The heterogeneity in terms of ethnic, sectarian and town of origin identities became apparent due to a number of developments. The main reason behind this was the emergence of conflicts between different groups as a result of the radical politicization of *gecekondu* space, caused by the occurance of two factors. First of all, the emergence of a strong form of favoritism (particilik), which means the unequal municipal treatment of different regions in *gecekondu* neighborhoods according to their political affiliation in terms of providing services, emerged during that period in a strong form. The municipal services such as the construction of roads, parks, schools, water, electricity and sewer systems and provision of transportation facilities constituted a bargaining power for both *gecekondu* dwellers and local politicians before the elections. An intimate dialogue between political candidates and *gecekondu* inhabitants during those years was evident.

The second important underlying reason behind the politicization of these settlements was the economic and political depression in the country. During the period between 1950 and 1970, the Turkish economy experienced rapid growth, mainly financed by foreign debt. In 1970, in order to encourage investment, the Turkish lira was devalued. However, after the military memorandum in 1971, the government ended the policy of encouraging investments, citing budgetary deficit as the reason. The unsustainable economic policies, low productivity and inability to control inflation prepared the ground for the economic and political crises for the period between 1975 and 1980. The rate of marginal employment and unemployment was considerably higher in 1978 when compared to the late 1960s. The poverty experienced during those years facilitated the manipulation of *gecekondu* population by radical political parties. In other words, class based politics had triggered the radical politicization of sectarian, ethnic and *hemşehri* differences in *gecekondu* society. Economic instability was accompanied by political instability in such a way that this period witnessed the extreme polarization of politics between left and right wing ideologies. A number of extremist youth groups on the left, and the nationalist "Grey Wolves" and Islamic fundamentalists on the right experienced no difficulty in recruiting youngsters who had few or no career prospects due to the economic crisis.[139] The radical left considered the *Alevis* as natural allies due to the affinity with the egalitarian and humanistic nature of the *Alevi* belief system. *Alevism* adopted its religious content to the political ideology of socialism during

---

139 See Zürcher, Eric J. (1998), *Turkey: A Modern History*, London& New York: I.B. Tauris, p.276.

the 1970s and they mainly define themselves with the ethical framework of extreme-leftist politics.[140] "The fascist and religious extreme right, on the other hand, oriented their recruiting efforts on *Sünni* Muslims of the mixed regions by fanning their fear and hatred of the *Alevis*, thus, provoking violent incidents."[141] The sectarian difference of *Alevism* and *Sünnism* took on a political character. *Gecekondu* space, like other places in cities, was dominated by the violence and struggle going on between these two extreme groups. The main difference of *gecekondu* spaces from other neighborhoods in the city was the close encounters between different sectarian and *hemşehri* groups in the spontaneously developed space of these neighborhoods.

The politicization of the neighborhoods that began with favoritism gained impetus with the ongoing violent struggles between *Alevi* and *Sünni* groups. These two groups have defined their boundaries in neighborhoods and the entrance to each other's regions was severely forbidden. In that sense, neighborhood inhabitants had to take into consideration these regions in relation to their sectarian affinity when they determined their everyday routes in the neighborhood. These severe conflicts between the two groups took new forms in the following years. In the academic literature of the period, it is difficult to find information about this part of *gecekondu* history. However, the heterogeneous nature of *gecekondu* society in terms of ethnic, sectarian and hometown identities which constitutes the main subject matter in *gecekondu* research after 1980 will be dealt with the following pages. For now, it is possible to claim that the deconstruction process of the representation of *gecekondu* spaces under the homogenous category of "rural other" in public discourse started during the 1970s.

There are reasons other than the politicization of *gecekondu* space, which contributed to this deconstruction. Starting from the early 1970s, when forerunner migrants reached a certain level of economic and social well being, we can talk about the commercialization of *gecekondu* settlements. This was due to the continuous migratory flow to these settlements, and the perception of early *gecekondu* settlers, the land and house ownership as the main sources of economic gain. Therefore, every migrant, particularly the early comers, looked for ways to benefit from real estate at the utmost possible level. For example, if they had one house with three rooms they would rent one of the rooms to a newcomer migrant. They employed every chance to build a second *gecekondu* in their garden as a continuation of their houses, for the sake of renting it. This increased the number of tenants in *gecekondu* neighborhoods. The scarcity of urban land prevented newcomers

140 See Çamuroğlu (2000:15).

141 Bruinessen (1996:8).

from making profits out of land or houses as the early comers had done. As a result, there emerged important transformations in the estate system, which provided limited chances to newcomer migrants and put them economically in a very different position than the early comers.

When we look at the *gecekondu* settlements from a general perspective, differences emerged between them in terms of their location in the city. Şenyapılı pointed to three different forms of development among *gecekondu* settlements during that period.[142] The *gecekondu* settlements, which were smaller and closer to city center, because of the spread of the cities in all directions, were partly or wholly reconstructed as modern neighborhoods. They were composed of apartment houses constructed by big construction firms. The second type of *gecekondu* settlements was again closer to the city center but their land value was not so high. They were located close to lower middle class neighborhoods and had been abondened mostly by the first settlers who moved closer to the middle class neighborhoods. *Gecekondu* houses were rented to newcomer migrants in this type of neighborhood. Since land values in these areas did not show a remarkable increase, rent became the only income for the migrants. As mentioned in the introduction, the research setting has experienced such a spatialization where early-comer migrants had rented their houses to the new comer migrants and had moved to lower middle class suburban districts. The third group of *gecekondu* settlements were the ones located in the periphery of the cities. During the period between 1970 and 1980 the inhabitants of peripheral *gecekondu* settlements themselves had initiated the transformation in their settlements.[143] This is the reason why the peripheral settlements, similar to the neighborhood where I conducted research, had been transformed mostly by the initiative of their settlers. They shaped their space according to their own needs and this is the reason why *gecekondu* settlements mostly had their own centers close by.

To sum up, the *gecekondu* population attained an experiential maturity in the city environment during this period. Their numerical importance and accumulated knowledge on the political and economic relations in the city transformed them into participating agents with the capacity to divert the strategical acts of planning. The homogenous identity of the rural other has been deconstructed as a result of the visible differences among *gecekondu* population. This deconstruction is represented in public discourse with an overemphasis on the marginal qualities of *gecekondu* society. The ruthless political struggles between left wing and right wing groups through cultural

142 Şenyapılı (1998:311).

143 Ibid., p.311.

identities of *Alevism* and *Sünnism* had made visible the political significance of the.sectarian, ethnic and *hemşehri* differences within *gecekondu* society. What is more, differences emerged with respect to the economic and social positions of migrants due to the variations regarding timing of migration or diverse occupational chances that the urban context had provided.

2.3.2 Strategies of Planning: Dealing with an Immense Gecekondu Problem

The year 1963 marked the beginning of the planned period that witnessed the implementation of five-year development plans, and continued in the following four decades, resulting from the need for macro planning activity. These strategical plans included the systematic projections in order to allocate resources reasonably in service activities. In the five-year development plans, the issue of rapid urbanization and the accompanying problems were given priority. Nevertheless, the issues were considered in a formalistic manner in these plans, in congruence with the general strategical outlook on the *gecekondu* problem.

In the first five-year development plan covering the years between 1963 and 1967, the general objective was to prevent uneven regional development and to attain a balanced growth that would, in turn, end the problems of rapid urbanization in large cities. In order to tackle these objectives, investment schemes were prepared; although they could not be properly implemented. In relation to *gecekondu,* three principals were foreseen, improvement of *gecekondus'* condition where possible; demolition of the houses in bad condition with no hope of improvement and assured prevention of further *gecekondu* construction.[144] In accordance with these principles, *gecekondu* law number 775 was put into effect in 1966. This important law differs from other laws because it was the first formal acknowledgement of *gecekondu* settlements and it brought a detailed trilateral plan to the *gecekondu* issue, although it took the issue more as a physical problem than a social problem. It transferred the ownership rights of state lands within the municipal boundaries to municipalities in order to make them directly responsible from the application of the law. After this law was put into effect, the *gecekondu* settlements that were recognized as those regions to be improved were, in time, legalized and a degree of development was sustained. However, this law could not avoid the further proliferation of *gecekondu* settlements, although municipalities took the law seriously. The narratives in the research setting verified the intense supervisions of local government during this period. Most of them noted the year 1968 as the most difficult

---

144 Keleş (1993:394,395).

year for constructing a new *gecekondu*. The state officials, in some regions, including the research setting, pitched tents and stayed for days in order to observe and control the construction activities going on during the nights.

Up until 1983, the five-year plans were unable to develop a permanent solution to the *gecekondu* issue due to the amnesty laws implemented in 1976 and 1983. Though the rate of rural-to urban migration decreased as a result of the urban unrest experienced in large cities, the proliferation of *gecekondu* settlements could not be prevented because of populist concerns. In addition, the centralist system of local government lacked the financial resources to prepare or apply construction plans in an efficient way. In the big cities, the transition of municipalities from center-right parties to social democrat parties during the 1970s led to the opening of debate in the centralist system of local government. "*CHP* mayors were voted into office in major cities like Ankara, İstanbul and İzmir. These mayors started to play a more visible role both in national and local politics."[145] Particularly, the financial dependency of municipalities on the central government was severely criticized by *CHP* mayors but until 1980 they continued to be dependent for external funds on the state.[146] Populism dominated the scene of *gecekondu* policies in Turkey and planning activity and the laws could not be easily applied. The resistance of the *gecekondu* population and their ability to bargain constituted the basis of this populism. In 1969, the voter turnout of İstanbul and Ankara squatters was 51% and 59.7% respectively and considering the increasing importance of urban votes, squatter votes become critical for the political actors.[147] The political context of the country and the demands of the *gecekondulu* affected the nature of the measures taken by the politicians. With the shifting balance of the urban population in favor of the rural migrants during the 1960s and early 1970s in the largest cities with respect to old city dwellers, the attitudes and actions of politicians had become hesitant. From the 1960s onwards, the *gecekondu* laws provided amnesties for those *gecekondu* houses already constructed which sustained adequate living conditions. The aim of these official pardons was to include already constructed *gecekondus* within the context of construction plans, thereby preventing further proliferation of *gecekondu* settlements. However,

---

145 Ergüder, Üstün (1987), "Decentralization of Local Government and Turkish Political Culture" in Heper, M. (ed.), *Democracy and Local Government: İstanbul in the 1980s*, Eothen Press, p.16.

146 See Heper, Metin (1989), *Local Government in Turkey*, London: Routledge and also see Görmez, Kemal (1997). *Yerel Demokrasi ve Türkiye* (Local Democracy and Turkey), Ankara: Vadi Yayınları.

147 Danielson and Keleş (1985:101).

on the tactical level, these strategical decisions brought about unexpected consequences. Most of the times, amnesty laws prompted the rate of rural-to urban migration and, therefore, the expansion of *gecekondu* settlements. The vote-seeking political activities, in the form of amnesty laws, triggered the expectation of legalization on the part of the *gecekondu* dwellers.

During the second period, the legal context of urban planning policies provided a suitable context for the proliferation of *gecekondu* settlements in the peripheries of the cities. The construction, planning and amnesty laws were promulgated in order to consider *gecekondu* settlements within the formal context of urban planning. However, the apparent intention of the amnesty laws mentioned above had an implicit populist content and concern on the part of the political actors. These actors used the amnesty expectations of migrants at every election. In addition, the *gecekondulus* who had attained a certain experiential maturity in political and economic terms, became aware of their bargaining power vis-à-vis the politicians. These developments further triggered conflicting interests among the different actors in the strategical realm. The public and academic approach to *gecekondu* settlements continued to be shaped by modernist concerns; however, political actors started to attain a hesitant legal and planning approach to the development of *gecekondu* settlements. From time to time, the politicians internalized the modernist approach in relation to *gecekondu* settlements with regard to changing economic and political context.

2.3.3. Public and Academic Approach to Gecekondu Settlements Dominated by the Problem of Integration

Notwithstanding the increasing reciprocity between the tactical and strategical realm and the growing participatory agency of the migrants in local politics, the question of the integration of the *gecekondu* population within city life still dominated the academic approach during these years. The research conducted at the time was strongly influenced by the modernization approach as was the case in previous decades. In this respect, the studies seem to accept the "urban" and "urbanite" concepts in Western literature and applied theoretical urbanization templates to the Turkish experience of urbanization. The degree of integration was measured by applying such behavioral models to the lives of *gecekondu* inhabitants. In most cases, the ideal templates did not match the lives, cultural preferences of migrants and the way they integrated with the urban economic and political context in these studies. The *gecekondu* population was represented as marginal to the city in these studies.

The marginality debate dominated Latin American urbanization literature during the 1970s in a very similar sense with Turkish case. Basic premises of marginality debate can be listed as follows. First of all, urban poor is culturally marginal for the fact that they cannot develop urban cultural habits because of their traditional or rural cultural backgrounds and because of the poverty they experience in urban context. Secondly, they are politically marginal in the sense that they have a tendency to support radical and marginal groups. Their political marginality may be a consequence of another fact that they are ignorant about and indifferent to political issues in such an understanding. Third premise of the marginality debate is that they are economically marginal, since they cannot take part efficiently in the formal part of the economy.

It would be helpful to examine these marginality tenets with regard to the Turkish context. The premise of cultural marginality, as mentioned in the previous pages, can be accepted as a product of the modernization ideal and the ideal of attaining a uniform culture. However, the marginalization of the rural traditional background of Turkey reflected the ideological stance of the researchers, fed by the social distance between themselves and the migrants. Few considered the fact that urbanization, in the modern sense, was recent phenomenon in Turkish history. The culturalism dominating *gecekondu* literature in the 1970s did not present *gecekondu* "people's way of life" as the cultural preferences of these people, but as the product of poverty, ignorance or fatalism.[148] However, migrant communities hold to some of their cultural values, and construct peculiar and spontaneous spaces in the city in cultural terms. The insistence of migrants in protecting their "traditional" cultural practices is not related to their deprived position with regard to production and consumption in the economy as put forward by some studies[149]; nor is it related to poverty, leading to a fatalist and ignorant life style among migrants.[150] The choice of migrants, in cultural terms, finds its expression mostly in the way they produce and reproduce their everyday spaces. Moreover, this process does not, in reality, signify a "*gecekondu* way of life" that can be homogenously defined. This production and reproduction process is always kept alive via the cultural differences of *hemşehri* groups and the constant competition in the course of social spatialization. Paradoxically, the cultural marginality debate that attributes an otherness to *gecekondu* settlements, in some cases refers to a

---

148 See Türkdoğan, Orhan (1977), *Yoksulluk Kültürü* (Culture of Poverty), İstanbul: Dede Korkut Yayınları.

149 See Kartal, K. (1982), "Kentlileşme Sürecinde Toplumsal Değişme Odağı Olarak Ankara" (Ankara as the Focus of Social Transformation in Urbanization Process) in Türkiye Geliştirme Araştırmaları Vakfı (ed.), *Kentsel Bütünleşme*, Ankara: Vakıf Yayını, pp.123-163.

150 See Türkdoğan (1977) and Yasa (1970).

homogenous "*gecekondu* way of life", rural in character vis-à-vis urban culture and, in other cases, to conflict-ridden encounters between different cultural identities (sectarian, ethnic or hometown) and their threatening qualities for national unity. The definitions in public discourse can be accepted as context-dependent but have always been part of the "othering" process. Academic studies relying on a cultural marginality premise also have some structural and methodological problems. They make an abstract comparison between migrants and "native" city population without comparing them empirically. There is no research in the urbanization literature of that period that makes a real comparison between the everyday lives of the squatter inhabitants and middle class residents of the city.

A premise based on political marginality also needs to be critically evaluated. As noted in previous sections, because of the congruence of political developments, it is not possible to label *gecekondu* inhabitants as politically marginal. Rather, they tactically use local political means and participate in city politics more than the middle classes in the urban context. As some in-depth research indicate, migrants from the early periods of rural-to-urban migration are aware of how modern bureaucratic institutions function. In spite of the fact that they may depend on their acquaintances and communal networks[151] as buffer mechanisms in their encounters with the bureaucratic organizations, they have, at least, a theoretical knowledge about how to deal with the system case of difficulty.[152]

An economic marginality thesis supposes migrants' participation in the informal part of the economy as an indicator of their economic marginality. However, during those years the informal economy played a key role in the rapid industrialization and urbanization of the country. Particularly in the 1960s and 1970s, *gecekondu* inhabitants provided a supply of cheap and flexible labor sought by the economic system. As Şenyapılı[153] mentioned, in the consumption sphere of the economy they were considered as quite integrated in the general consumption patterns of the city population. Other than that, during the period between 1966 and 1975 early comer

---

151 Communal networks are most of the time organized along the lines of being from the same origin (province or village) in *gecekondu* settlements. These networks help migrants to solve their problems with state institutions or other problems related to city life.

152 See Heper, M. (1983), *Türkiye'de Kent Göçmeni ve Bürokratik Örgütler* (The Rural-to Urban Migrants in the City and Bureaucratic Organizations), İstanbul: Üçdal Neşriyat and also see Karpat (1976).

153 Şenyapılı, Tansı (1978), *Bütünleşmemiş Kentli Nüfus Sorunu* (The Problem of Uninetgrated Urban Population), Ankara: ODTÜ.

migrants were occupied in more organized jobs in service, construction and transportation sectors. The underemployment experienced in occupations such as street trader, porter, boot blacker, parking attendant and care taker had been handed over to the newcomer migrants. This fact is also supported by the life stories of the migrants in the research setting.

To sum up, the general approach that evaluates integration as the main problem of Turkish urbanization regards all the positive changes in the socio-economic position of migrants during the time spent in the city as positive steps taken on the way to modernization and integration with city life. The second or third generation migrants are thought to be fully urbanized when compared to the original migrants, according to the premises of these studies but this, in fact, hardly coincides with the actual situation.[154] As these studies are often quantitative in nature and target large samples, the indicators of integration rely on predetermined suppositions and models of ideal urban life and behavioral patterns (being well-informed political voters, going to the theater and movies regularly, having regular employment in the formal economy and dressing in certain ways, for example) in political, cultural and economic terms. However, urban behavioral patterns in these three aspects were defined with a strong reference to the "West" and to the modernization ideal of the Republican Regime. This line of thinking defines "urban" behavior in terms of select cultural ideas / ideals regarding civilized living, and disparages the inhabitants of squatter settlements as lacking in competence to assimilate "urban life styles".[155] Although these studies provide rich data about the socioeconomic conditions of migrant population (such as the income levels of the families, migration data, demographic data about families, data on the physical appearance of *gecekondu*, etc.) they were inadequate to develop a full insight into the everyday lives of the *gecekondu* inhabitants. Moreover, only a small number of exceptional studies considered the economic and political context of the country, and retain a relational approach to the experiences of *gecekondu* inhabitants with the agents of the city.

The most important response to the marginality debate in urbanization literature was posed by Janice Perlman (1976) in her book "The Myth of Marginality" as a product of the research she had conducted among Rio de Janerio's squatters. Bryan Roberts[156], one of the most quoted urban

---

154 See Sencer, Yakut (1979), *Türkiye'de Kentleşme: Bir Toplumsal ve Kültürel Değişme Süreci* (The Urbanization in Turkey: A Process of Social and Cultural Change), Ankara: Kültür Bakanlığı Yayınları, p.277 and also see Kartal (1982:141).

155 See Yasa (1970) and (1973); Türkdoğan (1977); Kartal (1982).

156 Roberts, Bryan (1978), *Cities of Peasants: The Political Economy of Urbanization*, London: Edward Arnold.

sociologists on Latin American urbanization, in a very similar vein, analyzed the underestimation on part of elitist approaches regarding the significance, the resourcefulness of the role of the poor and the extent to which they actively participate in urban economic and political life. Similar understandings began to dominate urban literature after the 1980s with the increasing social proximity between the researcher and the researched in Turkish literature and with the dominance of ethnographic methods. In the period between 1940 and 1980, the academic and public approaches have had overlapping concerns and were both shaped by elitist perspectives. However, in the coming decade we encounter a serious disjuncture between approaches of academic works and public debate regarding "*gecekondu* settlements."

2.4. 1980-Present Time: Further Commercialization of "*Gecekondu*" Space

2.4.1 Neoliberal Economic and Social Policies

After the military intervention of 1980, the restoring of the civil rule had taken a several years under the supervision of the National Security Council, the agent of the intervention. The political parties of the 1970s and their leaders were banned from politics for a ten-year period. A new constitution was implemented in 1981, which was restrictive in democratic rights and values compared to the 1960 constitution. New parties were founded under the close surveillance of National Security Council. There were three parties at that time, namely, The Nationalist Democracy Party led by retired General Sunalp on the right; Turgut Özal's Motherland Party occupying the center and Populist Party led by Necdet Calp, a former private secretary to İsmet İnönü, created with the aim of filling the vacuum left by the *CHP*.[157] Among these three parties, Motherland Party (*ANAP*) of Özal was distinguished by its liberal approach. The political strategy in the 1980s was defined by two seemingly contradicting tendencies. On the one hand, the military regime had taken significant repressive measures in the shape of depoliticization policies that aimed mainly at eliminating the radical politics and struggles taking place in the public context. Such an agenda included extremely far-reaching interventions in the everyday lives of the citizens in social and political terms, which will be detailed below. On the other hand, the same political strategy had given support to the first elected government who was a champion of economic liberalism, aiming at total integration with the world economy.[158]

---

157 Ahmad (1993:189).

158 Ayata, G. Ayşe (1997), "The Emergence of Identity Politics in Turkey," *New Perspectives on Turkey*, 17, p.59.

The reason for National Security Council's permitting the liberal policies of the government within the strictly controlled context of the strategical realm can be seen as closely related to the western support for Özal's party, especially among the financial circles.[159] The liberal anti-statist and anti-bureaucratic political agenda of Özal brought him the victory in 1983 election, which also reflected the covert preferences of those who were uncomfortable with the military's continuing interest and intervention in politics. However, the neoliberal agenda of Özal's party showed great correspondence and continuity with the depoliticization policies of the military regime in the aftermath of the intervention, which will be exemplified with the narratives of the migrants in the following chapters. As a matter of fact, neoliberal economic agenda including seemingly paradoxical events were completely compatible with the depoliticization policies of post coup d'état period. As Peck and Tickell[160] mention, by its definition, neoliberalism contains processes of liberal economic management together with a strongly interventionist social agenda, which can be seemingly paradoxical. However, liberal economic processes that are increasingly technocratic in form and therefore superficially "depoliticized" seem to have a complementary relation with a deeply interventionist agenda on certain selective social issues such as crime, immigration, welfare reform, urban order and surveillance, and community regeneration. In a similar vein, Turgut Özal put the liberal economic agenda in combination with interventionist social policies into effect after the 1983 election.

The neoliberal policies of *ANAP* initiated many transformations in the political, social and economic life of the country and the transformations in *gecekondu* space was not an exception to that general course of events. Particularly, the legitimacy crises *ANAP* had experienced just after the election and the populist policies connected with it had altered the socio-spatial context of *gecekondu* space dramatically. The first visible effect of these policies on the socio-spatial transformation of *gecekondu* settlement had become apparent with the populist concerns of the party, related to the 1984 municipal elections with the inclusion of two other political parties in the elections. The party had put *gecekondu* law number 2981 into effect eight months previously. The law had two important differences from other amnesty laws, which both underlay the reason for this law becoming a milestone in *gecekondu* history. The law permits the distribution of title deeds to *gecekondu* settlers who build on state owned lands, lands of foundations

159 Ahmad (1993:189).

160 Peck, Jamie and Tickell, Adam (2004) "Neoliberalizing Space", in Brenner, N. and Theodore, N. (eds.), *Spaces of Neoliberalism: Urban Restructuring in North America and Western Europe*, Blackwell Publishers, p.42.

or municipal lands with the condition that the settlers would pay for the land already appropriated. In order to be considered eligible for title deeds, the settlers were not permitted to own more than one house or a piece of land in any municipal district. The law acknowledges ownership rights of the settlers who built on privately owned lands with the condition of obtaining the consent of the landowner. As with other amnesty laws, the official objective of this law was to bring *gecekondu* settlements within the scope of official planning. However, the extensive coverage of the law to include almost all of *gecekondu* settlements within its provision, the timing, and the way it was applied denote the possible implicit populist concerns of the party. The law could not reach its objectives due to the way it had been carried out in quite parallel to its neoliberal agenda.

In addition to the above-mentioned law, a two-tiered municipal system was established in 1984 just a few weeks before the local elections. This system aiming at general decentralization in the administration of local governments was applied in İstanbul, Ankara and İzmir. A number of district municipalities were established within each of these metropolitan municipalities.[161] "*ANAP* thought devolving of powers to the localities would promote political democracy and make delivering of more and better services possible".[162] However, the internal impediments of this system when combined with the amnesty law mentioned above permitted greater freedom of maneuver for migrants' tactical action. Internal impediments of the system seem to be closely related to the financial dependency of district municipalities on the metropolitan municipality. A two-tiered municipal system was established with the aim of bringing, in theory, a general decentralization and autonomy to local government system; however, this objective could hardly be realized in practice. The financial dependency, when combined with the fiscal gaps and inefficiencies created conflicts and differences among various central government agencies that district mayors were in a position to exploit. As stated by Ayşe Öncü[163] in the absence of a substantial influx of public resources from the center, the "underpinnings of clientalistic power at the local level have come to rely increasingly upon selective implementation of regulatory and bureaucratic codes". These circumstances led to the emergence of incidents of repetitive budgeting, which enabled individuals to promote personal interests through clientalistic channels.[164]

161 Heper, Metin (1987), *Democracy and Local Government: İstanbul in the 1980s*, Eothen Press, p.4.

162 Heper, Metin (1989), *Local Government in Turkey*. London: Routledge, p.7.

163 Öncü, A. (1998), "The Politics of Urban Land Market in Turkey: 1950-1980", *International Journal of Urban and Regional Research*, 12(1), p.54.

164 Heper (1987:2).

If the political affiliations of the metropolitan and district mayors are conflicting, then the resources of the metropolitan municipality will be distributed unevenly in most cases, in fact, according to Üstün Ergüder[165] "even within the ranks of the same party, district mayors and the metropolitan mayor may potentially come into conflict in what one may call 'the rush to deliver services' and reap the ensuing political harvest." As stated by him[166], the metropolitan municipality tends to claim ownership of popular practices, and passes the responsibility for unpopular practices (such as *gecekondu* demolition) to the districts. Such responsibility issues trigger bias both on part of the district mayors and *gecekondulu*.

This problem is also supported by the research findings. The mayor of the Mamak district municipality is harshly criticized by the respondents for the repressive controls on illegal construction, incidents of demolition, and inability of him in dealing with the complex challenges of the localities. On the other hand, the Ankara metropolitan mayor, Melih Gökçek, is represented as successful local governor in almost all commentares due to the high profile aspects of the services he provides to Mamak in the form of the construction of highways or large recreational areas. In fact, both the district and metropolitan mayors are from the ranks of Islamist parties.

In contrast, there are also other cases where the district and metropolitan mayors who are within the ranks of the same party behave in harmony. In this case, the nepotistic policies to one ethnic or sectarian group in localities may easily be employed due to the backing of the central government. The main consequence of these developments has been the growing dominance of bias in local politics. Selective implementation of building, zoning, and planning codes seems to be the main instrument through which nepotistic policies have been carried out in neighborhoods.[167]

These nepotistic policies had a serious impact on the course of the legalization process in the 1980s, as has been gathered from the narratives of the migrants. In line with the neoliberal approach of *ANAP*, the designation of the *gecekondu* settlers entitled to be awarded title deeds, the associated bureaucratic process were passed to the initiative of suitably-qualified private technical offices under oath. The responsibility of a construction reform plan, an issue of major public concern, was left initiative of private offices, which led to some abuses. Until the application of the formal construction plans to these settlements, as an assurance of giving the official title deeds,

165 Ergüder, Ü. (1989), "Patterns of Authority", in Heper, M. (ed.), *Local Government in Turkey: Governing Greater İstanbul*, London: Routledge, p.37.

166 Ergüder (1989:43).

167 Öncü (1998:55).

these technical offices, working under oath, acted as intermediaries with the responsibility of determining those who would be awarded a special certificate called "tapu tahsis belgesi". After the consideration of these *gecekondu* buildings in the light of municipal construction and reform plans, these special certificates would be replaced with fully legal title deeds.

The law also contributed to the further commercialization of real estate system in *gecekondu* settlements, something far beyond the commercialization process beginning in the late 1970s. This was due to the fact that permission was given to the construction of four-story apartment buildings on the lands that were considered as appropriate to give title deeds. This law and the related successive laws issued during the 1983-1988 period, in general aimed at the rapid legalization of *gecekondu* settlements and the transformation of these settlements into organized districts composed of apartment blocks. The planning activity was not intended to take effect until after the initiation of the transformation activity, which was left to the internal dynamics of the settlements.[168] The neoliberal policies of legalization of *gecekondu* settlements and decentralization of the municipal system seemed to have an implicit objective of bringing spontaneous apartmentalization in these districts, which was intended to lower the costs of planning and construction of organized and planned apartment districts. It was assumed that permission given to construct four-story houses would activate the internal dynamics in local land market and attract the interest of construction firms by such means.

2.4.2 Changing Physical and Social Space of "Gecekondu" and Identity Politics

These developments had led to the whole alteration and intense commercialization of land ownership patterns as a totally new phenomenon.[169] First of all, two phases of *gecekondu* production, which were once considered as the indivisible parts of a single process, first, the appropriation of land and second, the construction have been separated. New actors had emerged as the appropriators of large areas of land. They divided these lands into small parcels, an illegal activity according to construction law, and sold them to recently arrived migrants. The land appropriation stage began to take on speculative, illegal and even mafia-like dimensions.[170] Secondly, the construction of houses

168 See Şenyapılı (1998:312).

169 See Erder, S. (1996), *Ümraniye: İstanbul'a Bir Kent Kondu* (Ümraniye: A City was Settled Upon İstanbul), İstanbul: İletişim Yayınları; (1997), *Kentsel Gerilim* (Urban Tension), Ankara: Um:ag. See also Işık, Oğuz, and Pınarcıoğlu, Melih (2001), *Nöbetleşe Yoksulluk* (Poverty in Turn), İstanbul: İletişim Yayınları.

170 Işık and Pınarcıoğlu (2001:160).

together with labor and construction materials became commercialized. A new construction sector emerged in which *gecekondu* became a commodity to a large extent. In this context, the newly emerging low-income settlements constructed as a consequence of such a process assumed a radically different appearance from the classical *gecekondu* settlements. The recently constructed illegal settlements in particularly İstanbul, İzmir, Diyarbakır and Mersin have quite distinct qualities from the classical *gecekondu* settlements. These settlement areas are subject to continuous and high rates of migration therefore experience a strong pressure on physical space. Keyder[171] defined the peripheral neighborhoods as being "distinguished by the unfinishedness of three and four-story buildings, constructed out of cheap concrete and brick and often lacking a final plastering, located haphazardly within what seem to be random settlement patterns." Another important peculiarity of these settlements originates from the lack of legal status of these settlements. In most of the classical *gecekondu* neighborhoods, however, particularly as a result of the amnesty laws established after 1980, *gecekondulus* obtained their legal title deeds and construction permits, with the exception of some houses built on topographically dangerous sites. As mentioned in the previous chapter, the designated research setting is such a mostly formalized classical *gecekondu* setting, which is quite different from these newly emerged illegal settlements in socio-spatial terms.

The socio-economic differences that were beginning to emerge in the previous decade were intensified as a result of the further commercialization of the construction process. Particularly in these newly emerging illegal settlements, but also in classical *gecekondu* settlements, the pattern in which a well-established migrant owns more than one house became widespread. As a result of this process, the number of tenants increased rapidly.[172] This economic difference between newly arrived and well-established migrants is accepted as two aspects of the same, dynamic process. Işık and Pınarcıoğlu (2001) refer to this process as "poverty in rotation." A new phase in the social spatialization of low-income settlements emerged as a result of the economic and social deprivations experienced in low-income settlements, combined with the strengthening of identity politics within the context of close encounters between different *hemşehri,* sectarian and ethnic groups in urban context.

---

171 Keyder, Çağlar (2005), "Globalization and Social Exclusion in İstanbul", *International Journal of Urban and Regional Research,* 29(1), p.127.

172 See Işık, O. and Pınarcıoğlu, M. (2001:162) and also Gökçe, B., and et al. (1993), *Gecekondularda Ailelerarası Geleneksel Dayanışmanın Çağdaş Organizasyonlara Dönüşümü* (The Development of Inter Family Traditional Solidarity Structure to Modern Organizations), Ankara: T.C. Başbakanlık Kadın ve Sosyal Hizmetler Müsteşarlığı Yayınları, p.160.

Identity politics has gained impetus by the mid-1980s as a reaction against both the increasing inequalities and intense depoliticization policies through locally repressive means and with the implementation of neoliberal policies. Identity politics has been materialized around three main axes.[173] First of all, the predominance given to the *Sünni* interpretation of Islam as a counerbalance to left-leaning politics of the military regime has given rise to Islamic "fundamentalism" within the context of depoliticization policies of state.

Kurdish uprisings in the Southeastern part of Turkey constitute the second axis of identity politics emerged after the 1980s. The fighting between state security forces and Kurdish separatists who aimed at founding an independent Kurdish state in the region have had serious impacts on the post-1980 form of rural-to-urban migration. Kurds of eastern and southeastern provinces predominated rural-to urban migration of 1990s. As stated by Keyder[174], compared to the previous flows of migration, push factors became more important than the pull factors in this recent type of migration. The newcomers' decision to migrate lacked any preparation, previous knowledge or social connection to the city. "This devastation is primarily due to the ethnic/separatist war, itself in part related, in various ways, to globalization and the collapse of national developmentalism."[175] The problem of urban ethnicity became visible with the increasing rate of migration of mainly Kurdish population to largest cities as a result of forced migration.[176] This had two-fold consequences for Kurdish migrants. On the one hand, the increasing interaction with other groups in the city context produced an increased awareness of their differences, leading to a strong sense of "them" and "us". On the other hand, they become aware of the existence of other Kurds from different regions of the country. This closeness in language and living conditions led them to perceive themselves as a group with collective identity, some developing a "minority consciousness".[177] The economic deprivations of Kurdish people in the urban context as a consequence of the forced migration constitute another source that strengthened identity definitions through class awareness. Rising anti-Kurdish sentiment in the public perception, sowed the seeds of conflict in mainly urban context, when combined with the restrictions of state on making the Kurdish ethnic identity explicit in public space.[178]

---

173 See Ayata, G. A. (1997), "The Emergence of Identity Politics in Turkey", *New Perspectives on Turkey*, 17, p.60.

174 Keyder (2005:131).

175 Ibid., p.131.

176 See Ayata (1997:62).

177 Ayata (1997:63).

178 Ibid., p.64.

Third axis of identity politics, which constitutes the main interest for the study, is closely related to the *Sünni*fication of state via the policies of military regime and the governments succeeding it. *Alevis*, as the second most populous, and therefore minority sect, have felt themselves threatened by this process. The close encounters between *Alevis* and *Sünnis* in the urban context, which took a conflict-ridden dimension with the struggles between leftist and rightist militants in the 1970s entered into a new phase after the 1980s. The main strategy of the state in privileging a *Sünni* interpretation of Islam has been Ministry of Education's granting secondary education privileges to religious schools. As stated by Ayata[179], the number of mosques and religious personnel multiplied and the State Directorate of Religious Affairs became an influential body employing almost 90,000 civil servants. Building mosques, even in *Alevi* villages, the opening up of Quran courses and compulsory high school religious courses in *Sünni* doctrine can be counted among these strategies, emphasizing the *Sünni* interpretation of Islam over the *Alevi*. Religious feeling in the mainly *Sünni* community increased parallel to the institutionalized religion. The spread of tariqat organizations and informal religious gatherings mainly among women particularly in low-income settlements have become widespread due to these supporting policies of the strategical realm.

The structural transformation in the municipal system regarding identity politics led to populism being applied at local rather than national level.[180] In this context, strategical realm's way of dealing with clashing identities at the local level had certain impacts on the nature of social encounters between different sectarian groups. Coalition governments defined the period between 1991 and 2002. According to Yaşlı[181], two influential phenomena at this time were "the sovereignty of state imposition around the axis of the rising Kurdish question and Pan-Islamism and the sovereignty of market mechanisms around the issues of stability measures and structural adjustment programs". With regard to religious influence, the victory of Welfare Party (*RP*), conservative Islamist party, in 1994 municipal elections in the largest of cities of Turkey weakened *Alevi* claims in social spatialization of low-income settlements as a consequence of the display of religious bias. The Justice and Development Party (*AKP*), which came to power alone in the 2002 national election constructed its election strategy on populist criticism of neoliberal strategy and artificially constructed identity politics mainly with regard

179 Ibid., p.69.

180 See Işık and Pınarcıoğlu (2001:166).

181 Yaşlı, Fatih (2004), "*AKP*, Muhafazakâr Demokrasi ve Yeni Sağ" (*AKP*, Conservative Democracy and New Right), *Birikim*, 180, p.43.

to Kurdish issue. The *AKP* was founded by advocates of change who, in opposition to the old ways of governing, initially formed the Islamist Virtue Party, the successor of the Welfare Party.[182] *AKP* took 35% of the votes in the elections and gained the capacity to change even the constitution owing to its overwhelming majority in the National Assembly.[183] The success of *AKP* "was considered by many to be a consequence of the rhetoric appropriated by the party in relation to the peripheral settlements"[184]. In addition to populist image making, the party's harsh criticism of free market thinking and its stress on social justice were largely responsible for strengthening their base.

However, the upcoming political strategies of *AKP* have reflected serious contrasts to their election discourse of social justice and poverty in recent years as with previous governments in Turkey. "Particularly, their ambition to integrate the country with the world economy and liberalize the economy, including the opening up of urban land to market mechanisms and the operation of global and national private market mechanisms seems remarkable."[185] As stated by Keyder[186], formerly populist politicians now mainly refer to market-mediated demands rather than the potential constituency of newcomer migrants. In that respect, they have shifted their allegiances from populist developmentalism to neighborhood improvement under the influence of capitalist logic according to the author. The display of neoliberal policies and the neglect of welfare and poverty policies in the late party program of *AKP* had devastating effects, especially for *Alevis* and more recent migrants, mainly Kurd in ethnic origin, the most disadvantaged two groups of the 1990s. By the 1980s, with the rising tide of social democracy in politics, *Alevis* became a politically significant force due to their inclination towards leftist politics.[187] However, as a consequence of the resurgence of *Sünni* politics mentioned above, *Alevis'* determining agency in local politics has declined to a large extent. The sectarian identities have been revitalized and repoliticized with reference to their religious content due to these processes. It is worth mentioning that the identity of *Alevism* has been

---

182 Welfare Party was banned from active politics by the decision of Constitutional Court in 1998 with claim of becoming the centre of activities contrary to the principles of secularism.

183 Yaşlı (2004:39).

184 Demirtaş, Neslihan and Şen, Seher (2007), "Varoş Identity: The Redefinition of Low-Income Settlements in Turkey", *Middle Eastern Studies*, 43(1), p.97.

185 See Demirtaş and Şen (2007:98).

186 Keyder (2005:130).

187 Ayata (1997:67).

reconstructed through initiatives of *Alevi* organizations and intellectuals in spite of the fact that they might not agree on a uniform representation of *Alevism* against strong *Sünni* domination. As will be made explicit in the empirical chapters, the competition between sectarian and *hemşehri* groups has become more embedded in the social spatialization process during this period due to the increasing reciprocity between strategical and tactical realms.

2.4.3 "Varoş" Replacing "Gecekondu": Representing Low-Income Settlements as "Illegal"

Beginning in the 1980s, due to the political and economic context mentioned above, the social and spatial distance between low-income settlements and upper-middle and upper class residential areas extended further. In congruence with the irreconcilable social distance between these two groups, "urban elite" perspective seems to be shaped by a stronger threat perception regarding low-income settlements. One important material reasons of such a negative approach towards low-income settlements is economic and social polarization, and its reflection in the landscape of cities as a serious spatial segregation. "New enclaves for the wealthier sections of the society emerge in the valuable periphery of the cities, while the urban poor reside in topographically the most unsuitable places for settlement."[188] The periphery of the cities, which once attracted little attention from the high-income classes, has become valuable estate as a consequence of the fact that suburban, socially hygienic life has become fashionable in the 1990s. That development has brought the two socially most dissociated groups together as the main interest groups of the peripheral land. The upper economic classes used their political and economic influence while the low-income groups have transformed their spaces and appropriated new lands in the valuable periphery through mafia-style organizations.[189] In some cases, the physical distance between these groups became minimized, which increased the degree of uneasiness on both sides. Inevitably, a new social and legal imaginary that has been internalized by politicians and residents alike emerged in these years.[190] "Hitherto seen as poor people without resources to find adequate shelters, the migrants are now regarded as invaders of public property and beneficiaries of unfair privilege"[191]. This may be considered as

188 Demirtaş and Şen (2007:91).

189 See Şenyapılı (1998:311).

190 See Keyder (2005:130).

191 Ibid., p.130.

one important reason behind the appearance of discriminatory expressions in public debates on illegal settlements, particularly after the 1980s, showing *gecekondu* settlers as undeserving profiteers of the informal estate system as a consequence of changing construction processes and relations in illegal settlements. This labelling has gained a new form in the 1990s with the widespread usage of the term varoş either in place of or accompanying *gecekondu*. *Varoş* as a concept has a stronger emphasis on culturally pejorative, illegal, criminal or underground aspects of these settlements when compared to the concept *gecekondu*. The negative public perception of peripheral settlements that have been reshaped around the new term varoş and the novel usage of *gecekondu*, loaded with new meanings have a strong homogenizing effect. They dissolve all the differences between old classical *gecekondu* settlements that have been legalized to a large extent and the new illegal neighborhoods which emerged in the largest cities of Turkey after the 1980s. Therefore, varoş and the representations identified with the term that mainly define culturally inferior and criminal lifestyles with reference to the newly emerging illegal settlements have started to define all low-income settlements including the already legalized classical *gecekondu* districts. The representations of *gecekondu* and varoş highlight the illegality of peripheral neighborhoods in public debate in a number of ways.

First and foremost, peripheral settlements are portrayed as illegal, primarily referring to the illegal construction network including outlaw actors such as land mafia and speculators. The second source of illegality defined discursively has references to the tension-laden encounters between different ethnic and sectarian groups in low-income settlements. In the third sense, there emerged a portrayal of illegality mainly with reference to the criminal nature of these places as a consequence of their sheltering poor people.

The illegality in the first sense is related to the processes of the construction of peripheral settlements. In this realm, public discussions give references mainly to the concept *gecekondu* rather than varoş in defining low-income settlements. *Gecekondu* is mainly identified with illegal construction, its problems and demolition incidents. Some stories implicitly make allusion to represent *gecekondu* settlers as undeserving beneficiaries of urban land and at the same time aggressive actors in the struggle against demolition incidents[192]. It is considered as the space for providing illegal income for a network of outlaw actors and its inhabitants. It is interesting to note that in parallel to the policy changes on part of *AKP*, their approach and actions regarding low-income settlements have also reflected a certain transformation from rhetoric of social justice and social deprivation to a discourse of virtual hostility towards

192 See Hürriyet [İstanbul daily], 1 August 2002.

illegal housing and settlement.[193] The classical elitist approach to *gecekondu* settlements, giving references to the importance of migration control to large cities or to *gecekondu* development as an important impediment to civilization seems to have been appropriated by Erdoğan in a number of public speeches.[194] He refers to uncontrolled rural-to-urban migration as presenting a serious security threat to the order of the cities, and for this reason he envisioned visa (nakil il-muhaberi) application during emigration to large cities as essential. Paradoxically, this change of approach disappeared together with intense media interest in Erdoğan and his family's visits to poor families in *gecekondu* neighborhoods, particularly during Ramadan.[195] This perception change on part of the party has a strong connection with their policies regarding EU integration as well as with the neoliberal urban policies they put into operation. In spite of the breakdown of the alliance between populist policies of governments and *gecekondulu* due to the nature of new populism[196], middle class spectators still perceive *gecekondulu* as the main beneficiaries of the rent of urban land.[197] However, as mentioned above, illegal construction is presently not tolerated by the government and demolition decisions are strictly enforced in the face of serious protests and resistance of informal settlers.

Illegality in the second sense is related to the potential of low-income settlements to create a tension between different ethnic and sectarian groups encounter each other in confined areas. Due to the dense settlement patterns and the ineffectiveness of old migration and settlement processes, the migrants are unable to find the opportunity to settle near their fellow countrymen, as in the case of classical *gecekondu* settlements up until the late 1970s.[198] Identity politics had also contributed to public approach of the conflict-ridden encounters in these settlements. Particularly in the newly emerged illegal settlements, broader affiliations, in other words, ethnic and sectarian identities, carry greater importance than classical identity formations around small *hemşehri* communities.[199] The existence of such polarization based on

193 Demirtaş and Şen (2007:98).

194 See Sabah [İstanbul daily], 1 November 2003 and also Zaman [İstanbul daily, Islamic in orientation], 27 September 2005.

195 Demirtaş and Şen (2007:99).

196 New populism due to its compromise with global and national capitalists regarding urban land prefers to break an important part of its alliance with *gecekondu* settlers, since the politicians no longer want to offer state land and services to them.

197 See Demirtaş and Şen (2007:99).

198 See Tok (1999:83).

199 See Ayata, Ayşe-Güneş (1990-1991), "Gecekondularda Kimlik Sorunu, Dayanışma Örüntüleri ve *Hemşehrilik*", *Toplum veBilim*, 51/52, pp.89-101.

identity in such close quarters seems to contribute to the claims regarding the illegitimacy of these places. The usage of the term varoş is mainly encountered in such a context signifying the socially isolated, closed communities and their potential threat to peace and national unity. Particularly, the Gazi neighbourhood affair and the incidents that took place on Labor Day, May 1, 1996 triggered the anxiety of "urbanites". Gazi neighborhood, an unplanned low-income settlement in İstanbul, was thought to have a population of around 100,000 when it appeared at the forefront of public debate on 12 March 1995. Most of the neighborhood's settlers had migrated from southeastern Anatolia and they were *Alevi* and Kurd in origin. A small group of people whose identities were not determined, attacked four coffee houses mainly used by *Alevis* and a *Cemevi* with machine guns and killed a nuber of people. As a result, *Alevis* living in the neighborhood, together with other *Alevis* and leftist groups from İstanbul and other regions of Turkey, while protesting the event, also protested the indifferent attitude of the police forces leading to violence with security forces. These struggles escalated into serious conflicts between the two sides. The harsh response of police forces in repressing the protests contributed to the warlike nature of the event. The event paved the way for a renewed public perception of *gecekondu* settlements; the Gazi affair was portrayed in newspapers predominantly as a result of the explosive growth of *gecekondu*.[200] Although articles appeared to avoid using provocative remarks, they implicitly identified the neighborhood with violence by referring to the inherent qualities of *gecekondu*, and the history, social, cultural and economic structure of the neighborhood was defined as inherently carrying the potential for such conflict, crime and violence.

The second event that triggered the perception of peripheral districts as inhabiting tension-laden relations took place on May 1, 1996. Thousands of workers from *gecekondu* neighborhoods, alongside with many leftist organizations and militants of illegal organizations, initiated a march to Kadıköy, an important middle class district in İstanbul, in order to celebrate Labor Day. However, the celebrations deteriorated into a violent struggle between demonstrators and police forces towards the end of the meeting. A number of the demonstrators attacked the police, nearby shops, and banks located on either side of the road. After these two events, the term varoş has been used more frequently in public discourse, and the news and editorials on political conflicts and social tension in peripheral spaces often make use of this term in preference to *gecekondu*.

The third source of illegality is connected with the poor living conditions in the newly emerged low-income settlements and their potential for

200 See Hürriyet: 15 March 1995 and also Sabah: 19 March 1995.

poverty-induced criminality. Crime is often discussed with reference to varoş youth, who are portrayed as having a tendency to criminality since they are affected most by low standart of living in the city.[201] The increasing rate of crime in large cities is explained almost always with reference to the numerical superiority of peripheral settlers, and the associated dangers.[202] The commonly used verb in conjunction with varoş in the public debate in the 2000s is "invasion", referring to invasion of both city space and culture. *Gecekondu* retains its meaning mainly as a spatial signifier of peripheral life; rather than defining the peripheral space particularly, varoş began to be used to refer to a way of life, a culture predominantly carrying inferior and negative qualities, in contrast to high urban culture. A way of life associated mainly with crime, violence, undesirable habits and undeserved earnings have dominated the definition of varoş in the 2000s.[203] Its strong cultural connotation makes the usage of the term less exact in identifying all inferior aspects in the definition of popular culture with the concept. The identification of the term with a particular way of life or certain attitudes, and therefore its transmission to cultural realm mostly brings about the emergence of a representation embracing racist overtones. Zeliha Etöz[204] points out the observable fact that *varoşlu* have been labeled as "black Turks" vis-à-vis "white Turks" in news and editorials beginning from the late 1990s. This discursive division between "black" and "white Turks", though encapsulating ethnic meanings and indicating a distinction between Kurdish[205] and Turkish identities, encompasses this distinction by classifying people whether they have appropriated urban traits and way of life or not. Therefore the racist overtones in the dominant discourse, besides referring to an ethnic distinction, introduce a novel form of urban racism by supposing a distinction between "urbanites" and "varoşlu". It is interesting to note that early migrants of *gecekondu* settlements perceive less prosperous later arrivals in a somewhat similar manner.

2.5 Conclusion

The "representation of *gecekondu* space" within a Lefebvrian understanding has never been a homogenous domain. The actors playing an active role

---

201 See Zaman: 16 February 2003.

202 See the editorial of Erol Katırcıoğlu, Radikal: 24 April 2004.

203 See Hürriyet: 31 July 2002.

204 Etöz, Zeliha (2000), "Varoş: Bir İstila Bir Tehdit", *Birikim*, 132, p.51.

205 Therefore it mainly refers to the rural-to-urban migrants from the eastern and southeastern parts of Turkey.

in representing *gecekondu* space through the acts of planning, developing policies, ordering and defining *gecekondu*, have adopted different subject positions such as modernists, populists, urban elitists, etc. Subject positions of representative actors change in relation to their interests vis-à-vis the shifting social and economic context of the country. These various attempts to represent the "other" of the modernist urbanization were realized also in accordance with the changing socio-spatial context of *gecekondu* settlements over time.

As discussed in the introductory chapter, the strategical approach of the politicians, state bureaucrats, urban planners and academics in dealing with *gecekondu* space, and tactical acts of migrants exhibit an intertwined character. The dialectical interaction between the "representations of space" (conceived space) and the "representational space" (lived space) in Lefebvrian sense had led to the emergence of certain "spatial practices" as the third component of spatial triad. In other words, the dialectical interaction between representing *gecekondu* space and representational *gecekondu* space as lived through the acts and conceptualizations of the *gecekondulu* in everyday life have led to the emergence of certain spatial practices, which have not been always coincided with the suppositions and predictions of strategical realm.

What further complicates the picture is the continuous interaction between the various actors in opposing realms. It is necessary to remember that these actors have different subject and power positions in each realm of the spatial triad. Lefebvre and De Certeau overemphasize the determining capacity and power of strategical realm in producing a modernist space in comparison to the capabilities of tactical realm. Furthermore, Lefebvre and De Certeau mainly attribute homogeneity to these two realms. However, as examined in depth in this chapter, the growing experiential and political power of migrants in the city context increased the reciprocity between strategical and tactical realms. This reciprocity had further fragmented the strategical realm, which had never formed a totally homogenous whole, even in the early Republican period. The strategical realm had been fragmented around different interests and subject positions in relation to the changing socio-spatial composition of *gecekondu* space.

In the first period of *gecekondu* history, the political actors, Republican bureaucrats, journalists and academics had internalized quasi-uniform discourse about *gecekondu* space that was shaped around the modernist and elitist perception of the time. In the second period, unlike the first, the political actors had internalized a more hesitant subject position with regard to *gecekondulus* around populist concerns. In this period, *gecekondu* settlements became more visible and had rapidly proliferated in the periphery of the cities. Moreover, migrants gained capacity and knowledge to exploit the favoritism which dominated Turkish politics (loopholes of the strategical realm) to their own ends. The academic approach during this period began to present the system related causes of *gecekondu* growth but continued to emphasize the integration problem of *gecekondulu* within a modernist framework as the most important urban problem. As for the third period, with the increasing visibility of ethnic, sectarian and *hemşehri* heterogeneity in low-income settlements and the increasing effect of identity politics, perception of threat, centred on illegal aspects of *gecekondu* space dominates the public discourse. The middle class spectators represent *gecekondu* space as an illegal space in terms of its harbouring of conflict-ridden ethnic encounters, poverty and capitalistic, mafia-like construction processes. In the post-1980 period, however, the academic approach to low-income settlements, to a large extent, has abandoned its modernist, elitist conceptual schema. Recently, in-depth ethnographic methods have gained validity in understanding the dynamics of low-income settlements. Political actors' approach to low-income settlements had taken more hesitant forms with their squeeze between the demands of global capital, urban middle and upper middle class interests and low-income settlements.

CHAPTER 3 - The Emergence of Mamak *Gecekondu* Region in Ankara, the Heart of a Westernization Project

3.1 An Overview of Ankara's Urban History

Republican Regime's attempt to construct a modern capital in Ankara after the independence war is often depicted in the writings of many historians as an endeavor to build totally a new and modern space from scratch. In these writings, pre-Republican Ankara is portrayed as a small town carrying the traces of Independence War. The early urban experiences of the town are ignored, with a prominence given to its construction as a Western and modern urban space by the Republican Regime. However, according to archeological evidence, the history of this small town dates back to 3000 B.C., even before the Hittites. Its acknowledgement as an urban settlement dates back to as early as Phrygian civilization.[206]

Under the Roman and Byzantine rules, Ankara employed a predominant position due to its critical geographical location. Particularly, under the Byzantine rule, the city was given utmost importance in relation to its location on one of the most important long-distance interregional trade routes connecting Byzantium (later Constantinople) to the Eastern frontiers.[207] Under the Seljuk rule, the city occupied only a minor importance. However, the *Ahilik* system, a guild organization of the time, which was developed as a result of the need to unite all the craftsmen and artisans of Turkic communities migrating from Central Asia to Anatolia, was largely developed in Ankara.[208] This increased the commercial importance of the city, and also led to the development of the social organization in the city contributing to its urban characteristics.

Ankara had retained this importance as a trade and production center during the Ottoman rule, specializing in woolen fabrics and leather production. Particularly, in the first quarter of the 20th century, Ankara became an important center for the production of raw and semi-finished materials, as with many other Anatolian cities.[209] Angora wool was the most important of these products, and it ascribed a privileged economic status to the city. This process had modernized local production forms and

206 Balcıoğlu, Emin. M. (1993), "From Ancyra to Ankara: A Saga 3000 Years Long" in *Identity Marginality Space: Pre-Symposium Lectures and Panel Discussions*, METU Architecture Press, p.62.

207 See Balcıoğlu (1993:63).

208 Ibid., p.63.

209 Ibid., p.65.

enriched certain classes of non-Muslim origin such as Greeks, Armenians and Jews. At the beginning of the 20th century the class structure of Ankara was composed of these non-Muslim merchant groups, government officials, impoverished craftsmen and petty merchants. With the loss of the monopoly of the Angora wool and the shifting status within the changing context of the European economy, the lively economic atmosphere of the town started to fade a way particularly in the later decades of the Ottoman era.[210] Towards the end of War of Independence, as was the case in many other towns of Anatolia, non-Muslim merchant population started to leave the country and this led to the gradual transition of commercial activities to Turkish population.

Soon after the War of Independence, both because of the economic consequences of the war and the structural changes in the economic life of the city due this emigration, Ankara had experienced its most stagnant period in economic and social terms. The Turkish population of the city was largely engaged in rural activities (agriculture and livestock), at the time and the social organization of the city had lost its previous sophisticated nature.

To give a brief account of the pre-Republican history of the city seems important for one major reason. The social and economic context of the city was largely ignored within the framework of the Republican Regime's planning experience, as if it were an empty space devoid of already existing "urban traits" and relations. The new lifestyle, and the physical space constructed accordingly, were imposed upon the native population, which led to their alienation to a large extent. This seems to be closely related to the inherent qualities of modernist planning strategy of Republican Regime that inevitably brought about a conflict between the realm of "representation of space" (perceived space) and "representational space" (lived space) in the Lefebvrian sense. The social duality between "natives" and the "outsiders" in the early years of the construction of Ankara as a capital city reflected the intrinsic deficiencies of planning activity founded upon this irreconcilability between two realms of the space. Moreover, the social duality led to a split in the "social" and "physical" space of the city and the social spatialization following the initial planning years witnessed a process of contested use of space.

---

210 Çağlar, Özlem (2001), "The Atatürk Boulevard as a Locus of the Modernization Project of the Republic", Unpublished master's thesis, Ankara: Middle East Technical University, p.38.

3.2. The Historical Reasons behind the Strategy of Constructing a "Modern" Space in Ankara

The construction of the new capital city, small Anatolian town with a population of 20.000 involved many contradictory processes. The main reason behind the decision of moving the capital city from İstanbul to Ankara can be considered as the young nation's search for a new national identity outside İstanbul, the symbolic and the political center of Ottoman Empire. In common with the history of Ankara, the early planning experiences in newly established capital cities like Canberra, Brasilia or Islamabad, served the purpose of accelerating the process of nation building by constructing a new physical environment reflecting the new life style of the nation.[211] When the historical and symbolic importance of İstanbul is taken into account, this decision especially indicates a radical break from the past. The Republican Regime's objective of constructing a modern capital and developing the region may also have emerged from a need to construct an alternative to İstanbul city both economically and symbolically.

Ankara, with its central position in Anatolia and its proximity to other Anatolian towns long neglected by the semi-colonial economic and political policies of Ottoman state, was regarded as having the greatest potential for holding the country together.[212] Rather than paving the way to the intensification of inter-regional communication and the development of trade between the towns of Anatolia, the economic policies of the time were mainly aimed at improving communication between İstanbul and European countries mostly. This triggered the uneven regional development in favor of İstanbul and, in fact weakened the economic and communicative ties between İstanbul and other Anatolian towns. Therefore Ankara's favorable geopolitical position during the War of Independence as the terminus of the railway that had sustained transportation and communication between İstanbul and Anatolia since 1892, constituted another reason for its selection as the capital city of the young Republic.[213]

---

211 See Tankut, Gönül (1990), *Bir Başkentin İmarı: Ankara 1929-1939* (The Constrcution of a Capital City: Ankara 1929-1939), Ankara: Orta Doğu Teknik Üniversitesi Yayınları, p.28.

212 Altaban, Ömer (1987), "Ankara Kentsel Alanının Doğal Çevrede Yayılımı 1923-1985" (The Spread of Ankara's Urban Area in Natural Environment 1923-1985) in Tekeli. İ. et al., (eds.), *Ankara: 1985'den 2015'e*, Ankara Büyükşehir Belediyesi, p.42.

213 Erim, Aydan (1993), "Verbal Snapshots as Urban History" in *Identity Marginality Space: Pre-Symposium Lectures and Panel Discussions*, METU Architecture Press, p.76.

Moreover, considering the importance young Republican Regime had given to the ideal of secularism, Ankara had to be constructed as a strong secular alternative to İstanbul, which used to symbolize the religious importance of the Caliphate.

The construction of Ankara as a modernist practice and ideal integrates two different attitudes towards the Western world. As mentioned above, it was important to construct a capital city and a civilized model of life for the young nation far away from İstanbul, which was associated at the time with Sultanate regime and its dependencies on Western imperialist powers. At the same time, as a nation that turned its face to the western civilization, this new city needed the recognition from the civilized Western world as the center of the new nation, which was in the process of finding its place in the modern world.

Yet, the political reservations of European states in recognizing Ankara in place of İstanbul finds its expression in their hesitancy of moving their embassies to this "desperate" and "primitive" eastern town. It is easy to find the expression of this hesitancy in the memories of the early visitors to Ankara. Foreign visitors and officials described life conditions in Ankara as hard and unbearable. D.V. Mikusch, a foreign official, explained the alienating aspects of life for the foreigners in Ankara as follows: "life dozes in a slowness that is peculiar to Eastern people in Ankara"[214]. However, for the students coming from the other towns of Anatolia to Ankara for education, the town symbolized civilization with its small restaurants developed from traditional kitchens, schools built from hewn stones and wide streets with dirt and mud.[215] Falih Rıfkı Atay, a bureaucrat of the young Republic, defined Ankara as a symbol for the whole country outside the boundaries of İstanbul.

The European states led by England, campaigned against the idea of recognizing Ankara as the capital city.[216] In relation to this, another strong rationale emerged for constructing a modern infrastructure, in addition to building a modern national life style, namely, satisfying the practical needs of foreign bureaucrats. This whole project can therefore be summed up as an attempt to westernize the city both as a move away from the western world (in to Anatolia) and towards it (in accommodating Western diplomats).

---

214 Evren, Burçak (1998), *20'li Yılların Bozkır Kasabası Ankara* (Ankara: The Steppe Town of the 1920s), İstanbul: AD Kitapçılık, p.49.

215 Ibid., p.44.

216 Şimşir, Bilal N. (1988), *Ankara: Bir Başkentin Doğuşu* (Ankara: The Birth of a Capital City), Ankara: Bilgi, p.266, 277.

There were four main social groups in Ankara in the early 1920s. Besides the "natives" of the city, there were three major groups namely, migrants including students coming from smaller Anatolian towns, foreign officials, and Turkish bureaucrats. There was an emerging demand for accommodation for these three groups. Foreign officials mostly were on temporary visits returning to their countries on completion of their work. In this sense, the first modern buildings were the service providers such as hotels and restaurants.[217] These modern buildings also provided a space for the western style of urban entertainment that the Republican Regime imposed as the life style of the new nation. As an example, Ankara Palas, the first such hotel of the city, hosted many Western style dance parties given by Mustafa Kemal Atatürk reflecting the new life style of the nation. As stated by Baydar[218], the interior space of Ankara Palas was produced by using mechanisms of control, the observation of those who were denied entry, and the power of these outsiders' imagination. "For the privileged inhabitants of Ankara, dressed elegantly in accordance with the fashion pages of local newspapers that brought the latest news from Paris, and having adopted appropriate codes of behavior based on how-to books on "modern manners", ballroom dancing was part of a performance that they rehearsed only ambivalently".[219] At the opening night of the Ankara Palas, a ball was organized. Karaosmanoğlu, in his novel "Ankara", refers to this reception, portraying the social and cultural distance between the high state officials and the local population who constituted the fourth social group in Ankara at that time. As described by him,[220] in their European-style dresses and accessories, the guests arrive. The luxurious cars are all parked in front of the hotel. The native population stares at the ball crowd with wonder and astonishment having very little idea of what is going on inside the hotel.

From the beginning the social distance and life style differences between the natives of Ankara and the newcomers to the city decreased the possibilities of close social encounters. In cases where the natives and newcomers needed to communicate with each other, the differences in their dialects caused uneasiness. As stated by Burçak Evren[221], referring to the anecdotes of Falih Rıfkı Atay, the local people perceived the newcomers to the city as strangers and did not want to talk to them. Even if they had to talk, they never understood each other because of the differences in dialect.

---

217 See Evren (1998).

218 Baydar (1997:197).

219 Ibid., p.198.

220 Karaosmanoğlu, Y. Kadri (2000), *Ankara*, İstanbul: İletişim, p.116, 117.

221 Evren, Burçak (1998), *20'li Yılların Bozkır Kasabası Ankara* (Ankara: The Steppe Town of the 1920s), İstanbul: AD Kitapçılık, p.67.

As the main disciples of the new regime and its modernist framework Turkish bureaucrats in Ankara city had a mission to carry on the Western life style to the greatest possible extent.[222] However, adaption to modern daily habits, in the context of male-female relations made entertainment and consumption very difficult, even for the state officials and bureaucrats for social and economic reasons. Thus, it is understandable that these new habits were accepted as ridiculous and intolerable by the natives of Ankara who had been shaped by the qualities of Anatolian culture and the prohibitions of Islam. Even witnessing these changes was unbearable for some of the natives of the city.[223]

The social and cultural distance between the local people and the "Western strangers" and the neglect of the city's social context in the modernist planning strategy of the Republican Regime constituted, at the same time, its main weaknesses. As mentioned in the introductory chapter, the "production of space" in the Lefebvrian formulation emphasizes the power and capacity of the strategical realm to produce a totalizing space dominating all other spaces. The spontaneous nature of "other spaces" or the "daily urban reality" is ascribed limited power to divert urban space in Lefebvrian spatial triad. In cities like Ankara, which was planned as a modernist space reflective of a developing nation, the spontaneous spaces diverted and colonized the strategical modernist space. This was due to the neglect of the "lived space" by the agents of the "conceived space". The causes of such neglect will be discussed in detail below. It is important to note that this kind of neglect brings about, paradoxically, certain impediments for the ability of a strategical agency to produce a totalizing space.

3.3 The Weaknesses of the Planning Strategy

Strategical planning can be considered as the main component of any "modernization" ideal and practice. Ankara had become a platform for significant planning activity from the 1920s to the 1940s. After the initial construction of some hotels, restaurants and other emergent service-providing buildings, considered as the representatives of the First National Architectural Period, a large-scale planning activity was set in motion.[224] The products of this early period clearly reflect the ideal of Republican regime

---

222 See Şenol-Cantek (2006:46).

223 See Şenol-Cantek, Funda (2003), *Yaban lar ve Yerliler: Başkent Olma Sürecinde Ankara* (The Strangers and Natives: Ankara in the Process of Becoming a Capital City), İstanbul: İletişim.

224 See Evren (1998:174).

to reach a synthesis between East and West. The planning experiences of European cities constituted the main model for the planning of Ankara in the coming decade. In 1928, a competition was held among a small number of architects and Herman Jansen was selected as the creator and supervisor of the construction activity. His plan was set in motion legally in 1932. The plan included everything from the estimation of the city's population growth in 20 years time to the construction of every neighborhood, organized separately for different social groups in society, like high and middle ranked bureaucrats, workers, university teachers and students. Jansen's preliminary plan included the basic principles mentioned below:

> He proposed the citadel as the center of the city, surrounded with new districts. The traditional structure of the city would be renovated day by day with the connection of the new parts using the existing road patterns, and extending these roads to the new structure of the city. At the southern part of the city, Jansen proposed a governmental district in "Yenişehir" ["New city" was the planned city that would be constructed on the empty lands circulating the "old city"] newly constructed part of the city, consisting of grand plazas and monumental buildings connected to the Taşhan square [the center of the "old" or existing city] and the traditional fabric with a wide axis. He proposed low-density residential districts with houses of one or two stories placed in gardens in "Yenişehir".[225]

The structural and juridical arrangements were made in order to allocate a large amount of state financial resources to the service of this construction and to centralize the decision-making process. During this period, the main institution assembling the financial and political capacity to prepare and employ the plan was Ankara Construction Directorship. The application of the plan had a highly centralized administrative structure.

The premises of the plan, particularly in relation to the spatial development of the city on the main axis between two centers of Ankara, namely Ulus and Kızılay, was realized to a large extent. Today, these two centers still constitute the main connection axis of the city. However, some neighborhoods that were present in the actual plan could not be constructed. Furthermore, the planners could not control the unintended growth and construction held outside the planned boundaries of the city. There were many reasons for the partial realization of the plan.

---

225 Çağlar (2001:54).

The population growth and the rate of migration were high above the estimation of the planners. They could not construct the worker and university neighborhoods that were in the original plan since rural-to-urban migrants had built their *gecekondus* on the lands that had been legally expropriated by the state for the construction of these neighborhoods.[226] The growth of Ankara, the capital city, a city of attraction due to the concentration of bureaucratic posts and the many vacant positions, could not be well envisioned by the planners. At the beginning of the 1920s Ankara's population was estimated to be 20,000 people. This number reached 75,000 in 1927 and 290,000 in 1950. The development rate of Ankara surpassed the ratios in İstanbul and İzmir during the first decades of the Republican Regime. Between the years 1927 and 1950 the populations of İstanbul and Ankara increased to 42% and 48% respectively. For the migrants who came to the city, particularly those with no jobs or job prospects, there was little chance of owning or renting a house in the legal districts of the city. The real estate prices and rents were high during these years, even for the middle ranked state officials. Beginning from the early periods of planning activity, this unforeseen development led to the emergence of illegal settlement areas.

These settlements, perceived as temporary structures by state officials, met the accommodation needs of the migrants of that period. In this sense, they were regarded as providing a practical solution to the housing problem of migrants in the large cities of Turkey and the politicians were unwilling to demolish these neighborhoods for this reason.[227]

Land speculation constituted the main reason behind the tremendous value increases in land and housing prices. It constituted the second main impediment for the plan. During the period between the late 1920s and 1940s, the natives of Ankara and the state officials who owned land benefited from the speculative nature of the real estate market to the extent that their economic power, experience and relations allowed.[228] The native population of Ankara could not strategically evaluate the benefits of the land speculation or project the future value of their lands most of the time. An anecdote told by Aydan Erim[229], a witness of the period, demonstrates how

226 See Yavuz, Fehmi (1952), *Ankara'nın İmarı ve Şehirciliğimiz* (The Construction of Ankara and Our Urbanization Experience), Ankara Üniversitesi Siyasal Bilgiler Fakültesi, p.54.

227 Tankut, Gönül (1990), *Bir Başkentin İmarı: Ankara 1929-1939* (The Construction of a Capital City), Ankara: Orta Doğu Teknik Üniversitesi Yayınları, p.108.

228 See Yavuz (1952:63).

229 Erim, Aydan (1993), "Verbal Snapshots as Urban History" in *Identity Marginality Space: Pre-Symposium Lectures and Panel Discussions*, METU Architecture Press, p.76.

ignorant and naïve the native population of Ankara were in the decision-making process about their lands. A local man who had a large plot of land on today's Atatürk Boulevard, one of the symbols of the Republican's Regime's new Ankara, did not want to sell his land to another native and sold it to an "outsider". He explained his reluctance to sell this plot of land to his native friend in the following manner:

> "Look, my father cultivated that field, and after him I cultivated it and you know very well that it hasn't done any good to us, so you should feel yourself lucky. You are a local man. Why should I sell this useless plot of land to you?" which became gold, of course, in terms of land speculation in a few years time.[230]

Land speculation can also be accepted as the underlying cause of the housing shortage that was felt, mainly during the late 1930s and early 1940s. The price of land with construction permits was boosted during these years which led to the slowing down of construction activities in general. This also limited the constructors' ability in utilizing for construction the empty and available lands that constituted three fourths of the land with permits.[231] Erim's (son of a bureaucrat) childhood memories reflected the consequences of the housing shortage in the following manner:

> The rents were very high and landlords wanted lump cash payment before moving to the houses. It was like an auction... The one who offered the highest price as lump payment rented the house.[232]

House sharing with another family was an often-encountered fact for civil servants, even for those occupying important positions.

The negative reaction to Jansen plan and the unwillingness of the bureaucrats to cooperate with planners can be counted as another reason behind the failure of the plan. The upsurge of nationalism among both bureaucrats and planners of the time constituted the basis of the reactions against Jansen and his team. This was an attack against the dominance and delegated power of western planners in urban planning. Jansen was criticized harshly, particularly after the 1929 economic depression, on

---

230 See Erim (1993:76).

231 Altaban, Özcan (1998), "Cumhuriyet'in Kent Planlama Politikaları ve Ankara Deneyimi." in *75 Yılda Değişen Kent ve Mimarlık*, Tarih Vakfı Yayınları, p.48.

232 Erim (1993:77).

financial grounds. The criticism was due to a general reaction against the allocation of the nation's resources to foreign planners. Most parliamentarians emphasized the importance of using local human and material resources. Jansen's neglect of the lower-class neighborhoods of Ankara, namely the old city's lack of basic infrastructure, constituted the main point of criticism in parliament.[233] The preliminary plan of Jansen, whose basic principles are mentioned above, differed on some important points from his master plan. In the preliminary plan, he proposed the reconciliation of the old and new parts of the city; however, in the master plan he decided on a sharp separation of the old and new sections of the city.[234] This sharp distinction in physical space brought a fragmentation in the social spaces of the local people and outsiders, symbolizing the traditional and modern face of the city respectively. Yet, it is also worth mentioning the role of the parliamentarians in this significant change of plan. In the initial periods of the planning the young Republic and its parliamentarians agreed to leave the old city outside the scope of the plan because of the high land values.

Starting from the time that Jansen's plan was accepted, local and national efforts aimed at changing the plan emerged for different reasons and interests. Some high-ranking bureaucrats of young Republican Regime benefited most from the speculations mentioned above. Particularly after 1935, these speculative practices and decisions brought about organized and unorganized developments that were totally out of the plan's context. The Bahçelievler, Beşevler and Çankaya neighborhoods that are the important middle and upper-middle class districts of today's Ankara, were constructed as a result of such a process.[235] Jansen was also criticized for constructing houses and neighborhoods that were not suitable for a Turkish life style by the national architectes of the late 1930's. As a consequence of the criticisms and the difficulties in the application of the plan, Jansen was removed from the responsibility of implementing the plan in 1938. Whilst he was still in Ankara, Jansen asked for his signature to be removed from the plan. This indicated his dissatisfaction with the unintended developments that were outside the context of the plan.

To sum up, the main impediments behind the planning activity of Ankara can be categorized into two groups. The limitations intrinsic to any social engineering or modernization project constitute the first group of impediments. The modernization ideals and practices, as was the case in the construction of "modern" Ankara, did not take into account the social and

233 See Tankut (1990:68).

234 See Çağlar (2001:97).

235 Bademli (1987:107).

spatial context in which the model was conceived. To represent a modern space, the strategical realm needs to define its other. From a modernist perspective, most of the time, the "old" and "traditional" social and spatial structures of old Ankara were considered within the realm of this definition of "other". In order for a modernization discourse to become eternal and powerful this othering process needed to be kept alive. As pointed out by Banu Helvacıoğlu[236], unlike the modernization premises dominating the 1960s public discourse, the conceptual difficulty of calling Ankara a modern city was not a consequence of the presence of "backward" spaces in the form *gecekondu* settlements within it, but rather was due to the impossibility of realizing a hypermodern imagination. In this sense, the pre-Republican social and spatial context of Ankara had been given hardly any reference throughout the planning process. The social and cultural duality in the city, as well as the ongoing power of the modernization ideology on urban policy-making, contributed further to problems of the planning in the following decades. As stated by Balcıoğlu[237], at one stage during Nevzat Tandoğan's administration, in order to take a stroll on Atatürk Boulevard, a man had to wear a tie, and if he did not have one, he was not allowed to walk on the Boulevard. To create a new life style via the imposed modernist space of Ankara may have had to such extremes.

As mentioned above, the influence of land speculation, the high rate of migration and the reactions towards planning activities affected the course of events. The spontaneous social spatialization had, most of the time, preceded the plan. Therefore, the planning practices, in most cases, followed these unintended developments, even at the very beginning of the highly centralized and powerful first planning period of the Republican Regime. As a consequence of these developments, there appeared, in time, four "cities" in Ankara whose spatial characteristics and conditions of production differed significantly from each other.[238] The first of these cities was the "old city", or the historical city, that can be depicted as follows with reference to its structure in the 1920s:

> The urban structure was mainly comprised of housing districts upon the hill where the citadel is placed, and adjacent residential and commercial districts surrounding the citadel. Residential

236 Helvacıoğlu, Banu (2002), "Mekana Sığmayan Ankara'da Mekan Nasıl Yaşanmakta?" in Pultar, G., İncirlioğlu, E. and Akşit, B. (eds.), *Kültür ve Modernite*, İstanbul: Türkiye Kültür Araştırmaları, p.138.

237 Balcıoğlu (1993:78).

238 See Bademli (1993).

> districts comprised of an organic and compact layout with its dusty winding streets and simple mud-brick houses of one or two stories, which were mostly destroyed by the fire in 1915. The administrative center of Ankara was Taşhan Plaza located in the west end of the city, surrounded by wetlands on the West and connected to the station building in its south-west with the station avenue. Around the plaza there existed *Darül-Muallimin* (School of Teachers), the Building of *İttihat ve Terakki Cemiyeti* [The building would be used as the National Assembly Building, after Atatürk gathered the Assembly in Ankara], *Millet Bahçesi* (People's Garden), Taşhan, *Hükümet Binası* (House of Government), and the post office.[239]

Bademli labels "old Ankara" as the city that was neglected by the Republican Regime. Old Ankara was turned into the commercial center of the peripheral settlements in the following years. The second city was Yenişehir, the new city, was constructed at the south of the old Ankara city in the 1930's as the socio-spatial symbol of the modern life style during the initiation of the Republican Regime. This city among others displays most the Western urban structures and life style. In the late 1940s and early 1950s, a third city that Bademli calls the "instant city" surrounded both the old and the new city.[240] This instant city was composed mainly of *gecekondu* settlements. In the 1950s and 1960s the old quarters were destroyed in order to build new apartment houses which created another Ankara, named by Bademli as the "re-worked Ankara" or the fourth city.

3.4 Spontaneous Social Spatialization Altering the Course and Implementation of Planning

The emergence of "instant" Ankara, in Bademli's words, played a key role in the transformation of public space and the planning activities in the years to come. In this part, I will examine two incidents of spontaneous spatialization that had serious impacts on Ankara's public space by means of the development of peripheral settlements. First, I will elaborate on the spontaneous development of the routes of public transportation in Ankara. Second, I will examine the transforming meaning and functions attributed to the two city centers of Ankara as a consequence of the unintended social spatialization.

239 See Çağlar (2001:44).

240 See Bademli (1993:69).

In the 1920s, Ankara was mainly a city of pedestrians in accordance with its topography and size.[241] With the rapid increase in population towards the end of the 1930s, the vineyard houses that were located quite a distance from the city center and used on a seasonal basis, mainly by the families of bureaucrats, were transformed into permanent neighborhoods. Transportation became an important issue after that. The small entrepreneurs who gave transportation service with small buses called *"kaptı-kaçtı"* constituted the main agency of public transportation during these years.[242] Until 1935, the world economic depression prevented the municipality from taking efficient measures regarding transportation. With the initiation of municipal bus services in the 1960s, the rate of private initiatives in public transportation was reduced to 35%.[243] However, in the post Second World War years, the share of the municipality in public transportation facilities experienced a retrogression, both due to the low production ratio of vehicles and spare parts in the world and the fire that led to the demise of public buses in Ankara.[244] Between 1950 and 1954, Turkey increased its imports and this enlivened both public and private transportation facilities. However, when we look at the history of transportation and the share of private initiatives in public transportation in general, it is easy to observe that they played an essential role in Ankara's transportation. The main reason behind this can be put as the rapid population increase experienced particularly after the 1950s and the instant development of peripheral *gecekondu* settlements as a consequence of that. The official public transportation services, similar to the planning practices, followed the spontaneous social spatialization of the city. The municipal supply always remained insufficient.

The deficiencies of the municipal services were compensated by local private initiatives. With the rapid increase of population in Ankara, private minibuses increased in number and predominantly provided a service to *gecekondu* neighborhoods. The normal capacity of minibuses is 14 passengers; however they often take passengers standing beyond their permitted capacity. The emergence of private entrepreneurs servicing to middle and upper middle class districts followed a similar pattern to the minibuses servicing to *gecekondu* settlements. These vehicles, called *"dolmuş"* offered a service to these middle class neighborhoods for years.[245] *"Dolmuş"*

---

241 Tekeli, İlhan et al., (1987), *Ankara: 1985'den 2015'e* (Ankara: From 1985 to 2015), Ankara Büyükşehir Belediyesi, p.65.

242 Ibid., p.65.

243 Ibid., p.65.

244 Ibid., p.67

245 Tekeli (1987:68).

as a concept refers to the quality of the transportation service and these cars, similar to "shared taxis", do not start giving the service before they are fully occupied by passengers. Therefore, there are no standard time schedules for their arrival and departure. The *dolmuşes* that were providing services to middle class districts were different from other minibuses of the time. Unlike minibuses, they were normal cars with the capacity of up to four passengers. In the 1970s, the small entrepreneurs, with their minibuses and *dolmuşes* predominated the scene of city transportation.[246] They seemed to be symbols of the spontaneous aspects of spatialization in the city. Today, *dolmuşes*, as small cars, no longer offer a service in Ankara; rather, minibuses today give services to all neighborhoods, whether middle class or low-income settlements. The *dolmuş* was erased from the scene of public transportation from the early 1980s onwards. Minibuses, as well as private buses, affiliated with the municipality in administrative terms and public municipal buses still retained their share and importance in public transportation. The emergence of private buses corresponds to the 1980's when public transformation investments experienced a second phase of stagnation because of the economic stability programs. During the 1980s the composition of small entrepreneurs changed, and private public buses, compensating for both minibuses and municipal buses, entered the transportation sector as an important agent. Municipalities controlled and supervised private buses. For this reason, particularly in the early periods of their emergence, they were prevented by the municipality from offering a service to the most profitable routes.[247] However, in time they took a greater share of public transportation and have continued to give this service until the present. It is possible to observe that, with the neoliberal policies put into action in the 1980s, private initiatives have taken an essential part in public transportation. However, even from the 1950s onwards, private initiatives fulfilled essential roles in public transportation. Particularly, minibuses constituted the main means of transportation between peripheral settlements and the central districts.

To tackle the transportation history is crucial in one important respect. Particularly, in understanding the transformative capacity of *gecekondu* spaces on the formation of city space, the history of transportation provides important clues. The unintended and instant developments in the periphery of the city necessitated emergent provision of some services that could not be sustained by the municipality either financially or in planning. These local private initiatives that had emerged as a consequence of *gecekondu* formation determined the shaping of city space later. Planners

---

246 Ibid., p.68.

247 Ibid., p.69.

primarily took into consideration the already constructed minibus routes connecting peripheral and central districts while deciding on the official public transformation routes of the city.[248] These routes mainly extended from centers to peripheral settlements. The account of the shaping of the transportation routes exemplifies the impacts of the spontaneous social spatialization on the strategical realm via the spatial practices of the "lived space". The private initiatives of transportation, instantly developed in order to provide a service to *gecekondu* settlements, seem to have constituted the primary reference for planning attempts in the following period.

The transforming functions and meanings attributed to two of the city centers in Ankara, namely Ulus and Kızılay, constituted the second realm in which spontaneous spatialization was effective. Ulus was the center of "old city", serving as the only center in the early periods of the Republican Regime. Although there was an attempt by the Republican Regime to create a new center in the Yenişehir district, Ulus continued to be the center of state bureaucracy and a critical component of city space providing recreational and shopping facilities for many years. However, with the establishment[249] of Kızılay as the center of the new city, and with the impact of *gecekondu* development in the peripheral lands closer to Ulus rather than Kızılay in time, the symbolic and economic space of Ulus went through a significant transformation. It lost its attraction as a shopping center for middle and upper classes.

Beginning in the 1970s, Ulus started to be associated more with peripheral settlements rather than with the middle class districts. Kızılay served as the center of middle and upper middle classes until recently but today, Kızılay is also subject to transformation and has started to serve as the center for lower-middle classes and peripheral settlements like Ulus. In spite of the fact that Ulus protected its status as a commercial district until present, by the 1970s the nature of social spatialization changed in conformity with the introduction of new consumption patterns and customer profiles. Although the economic profile of the region has changed significantly, the commercial activities continued to survive and are concentrated more in Ulus than Kızılay. Futhermore, the bureaucratic posts constitute a greater share of the working population in Kızılay due to the positioning of state buildings in accordance with the premises of Jansen's plan.[250]

---

248 Ibid., p.70.

249 "Production" here denotes the constructed nature of Kızılay from nothing as a center of modern Ankara in Lefebvrian sense as mentioned in the first chapter.

250 See Akçura, Tuğrul (1971), *Ankara: Türkiye Cumhuriyeti'nin Başkenti Hakkında Monografik Bir Araştırma* (Ankara: A Monographic Research on the Capital City of Turkish Republic), Orta Doğu Teknik Üniversitesi Mimarlık Fakültesi Yayınları, p.101,109.

In the 1970s, with the increasing utilization of Ulus by *gecekondu* settlers, a functional spatial duality emerged in the district. Small retail and wholesale trading in some parts of the region provided a service to all the groups in the city, while in other parts of Ulus, the supply was peculiar to *gecekondu* settlers' needs and tastes.[251] Kızılay reflected a more homogenous picture during the 1970s and 1980s. This emerges from the fact that the middle and upper middle classes of Ankara mainly used Kızılay as their center.[252] Activities and services peculiar to middle class life dominated the scene of Kızılay during those years. According to the statistics of the year 1969, the ratio of hairdresser shops servicing women was 66.7% in Kızılay, and 33.3% in Ulus; the ratio of associations was 65. 8% in Kızılay and 34.8% in Ulus; the ratio of sheep and goats trading was 0% in Kızılay and 100% in Ulus and the ratio of psychiatrists was 54.3% in Kızılay and 45.7% in Ulus.[253]

In addition, the quality of the activities differed greatly in these two centers. For example, the associations of artisans, small merchants, and people coming from the same village were located in Ulus because of the physical proximity of Ulus with low-income settlements. The associations that can be considered as more institutionalized and modern, such as the Lawyers' Association, Women Associations, and Turk-German Friendship association, were located in Kızılay.[254]

The impact of peripheral settlements on the transformation of Ulus was obvious in Ankara. The most important reasons behind this was the physical proximity of Ulus to most of the peripheral settlements as stated above, and because of the railway connecting Mamak, one of the largest *gecekondu* districts with Ulus. Since the roads from the peripheral settlements all intersect in Ulus first, and the way to Kızılay is available only through Ulus, Ulus was transformed over time from the center of the Republican Regime's Ankara to the center of *gecekondu* settlements.[255] The boulevards, designed to distance the new city from the old city within the context of the "monumental narrative" as told by the Republican Regime, paradoxically functioned as " connections enabling the flow between" the centers of old city and new city, Ulus and Kızılay respectively.[256] The presence of the boulevards enabling the flow between Ulus and Kızılay constituted the main reason behind the recent shift in the functions of Kızılay and its provision of services to the inhabitants of the peripheral settlements who had become more mobile.

251 Ibid., p.121,122.

252 Ibid., p.126.

253 Ibid., p.126.

254 Akçura (1971:127).

255 Ibid., p.92.

256 See Baydar (1997:195).

3.5 "Post-Planning" Planning Strategies in Ankara

In 1957, the plan that became official, after Jansen's plan, was called the Yücel-Uybadin plan after the names of its planners. The political context of urban planning was transformed, to a large extent, in unison with the application of this plan in Ankara. During these years, *DP* gave İstanbul priority in terms of urban development and, therefore, investments were mainly shifted to İstanbul. Under the banner of new planning, big boulevards were constructed and legal expropriations were realized particularly in İstanbul.

Yücel-Uybadin's construction plan can be given as an important example of the planning tradition in Turkey, most of the time, follows spontaneous social spatialization. Yücel-Uybadin plan's main practice and objective was to restructure and order the unintended development and to prevent further *gecekondu* development, particularly in topographically unsuitable areas. Within the framework of the plan, they foresaw the organization of some unintended residential areas, including the Mamak region and villages close to it, like Üreğil and Kayaş.

Sharing a common fate with Jansen's plan, Yücel-Uybadin plan was unable to give an appropriate estimation on demographic development and the real estate and housing dynamics in Ankara. More than half of the population was living within the three-four kilometers-wide peripheral zones surrounding the city in the north and south mainly on the most inappropriate and highest regions of the city, where settlement was largely dangerous at the time. In the period between the years 1956 and 1969, low-income families who were unable to find suitable places for themselves in the formal settlement areas of the city, settled illegally in districts such as Balgat, Dikmen, Aşağı and Yukarı Öveçler, Abidinpaşa, Nato Highway, and Samsun Highway.

The plan could not include all these regions in its framework as a result of certain legal and practical limitations. The planners were constrained, legally, to realize planning only within the boundaries of the municipality[257] and, because of this, they were unable to plan for many of the unofficial settlements outside the boundaries of municipality. Moreover, they could not plan new residential districts as Jansen intended; rather they, concentrated on reforming the already constructed settlements. Due to its limitations, the plan envisioned a 750,000 populated, uni-centered, homogenous city and neglected the *gecekondu* settlements to a large extent.[258] With the

257 See Bademli (1987:107).

258 Ibid., p.107.

law issued in 1965, under the name of the "Story Ownership Law" (*Kat Mülkiyeti Yasası*), the construction of multiple owner apartment houses in building plots became possible.[259] This further prevented the applicability of laws associated with the Yücel-Uybadin plan, because it triggered rapid development of many new middle class neighborhoods. Since the plan ignored *gecekondu* neighborhoods to a large extent, the need for new and additional regional and partial construction plans and applications became inevitable.[260]

The transition of municipal posts to social democratic parties initiated an advantageous period for urban planning activities in Ankara. The planning activity conducted by the Ankara Regulative Planning Office foresaw the planned decentralization of the city in the western corridors in 1973. Between the years 1969 and 1984, the Ankara Metropolitan Planning Office emerged as the main agent of planning within the structure of the Construction and Housing Ministry. It sustained the coordination between the Ankara Construction Directorship and the Ankara Municipality. 3208 hectares of land were nationalized in order to construct *gecekondu* prevention regions in Ankara between the years 1965 and 1976.[261] Some parts of these nationalized lands were utilized effectively by implementing housing projects for low-income families while the migrants, for the sake of constructing *gecekondu* houses, appropriated others and this led to a failure of the projects in some regions.[262] In spite of the fact that prevention of many *gecekondu* regions was put into action during the period between 1965 and 1980, the number of *gecekondus* increased tremendously and 175,000 new additional *gecekondus* were constructed. This fact points to the failure of *gecekondu* prevention projects that aimed at transforming *gecekondu* areas into housing estates for the low-income population. These projects further triggered *gecekondu* construction by creating a supply of state owned lands in many regions that were appropriated for *gecekondu* construction due to the inconvenient and inefficient functioning of these projects. Migrants tactically utilized the delays in the application of these projects by colonizing these spaces.

In spite of the unintended developments in the city's spatial transformation during these years, the Ankara Regulative Planning Schema was acknowledged by urban scientists as bringing a dynamic and realistic approach to planning. This schema was prepared and applied during the

---

259 Altaban (1998:55).

260 Bademli (1987:109).

261 Türel, A. (1987), "Ankara'da Konut Yapım Süreçleri" in Ankara: 1985'den 2015'e, Tekeli, İ. et al., (eds.), Ankara Büyükşehir Belediyesi, p.61.

262 Ibid., p.61.

era, when social democratic parties, in general, took posts in municipalities. Ankara Metropolitan Planning Office worked in coordination with Ankara Construction Directorship, Ankara Municipality, universities and other public institutions. The plan successfully envisioned the development of Ankara. The population estimate of the plan, as an example, was considerably closer to the reality when compared to preceding planning experiences. The Ankara Metropolitan Regulative Plan was successful with many projects, like the Sincan *gecekondu* prevention area, Batıkent, Ankara Industrial District and similar developments.

However, the development of *gecekondu* regions could not be prevented. On the contrary, they gained further impetus during these years for a number of reasons. As mentioned in the previous chapter, *gecekondu* houses, particularly towards the end of 1970s, began to be constructed for investment purposes rather than for the purposes of providing shelter. Particularly, in neighborhoods that were legalized to a large extent and had transportation and infra-structural facilities, settlers were more inclined to build second *gecekondus* or additional stories. Vedat Dalokay accomplished many infra-structural improvements in *gecekondu* settlements in the 1970s. Some respondents of my research emphasize their happy memories of Dalokay's municipal period due to the projects that were carried out in coordination with the local people. Dalokay's municipal understanding and practice seemed to give utmost priority to the aim of providing services, beginning from the peripheral settlements to the city center.[263] The infra-structural services provided during this period contributed positively to the values of land and housing in *gecekondu* settlements. Loose control on *gecekondu* construction for different reasons, combined with the factor of increasing market values of these neighborhoods due the completion of services, led to the construction of new *gecekondus* for investment purposes during this period.[264] The approach to planning transformed, in a positive sense, populist policies, and applications were carried out with regard to *gecekondu* settlements. By 1984 the responsibility was given mainly to the metropolitan municipality to prepare and apply construction plans. Since then, the burden for municipalities, in terms of responsibility, has increased tremendously.

---

263 Altaban (1998:58).

264 Türel (1987:62).

3.6 The Remarkable Spontaneous Social Spatialization in the Periphery: The Development of the Mamak *Gecekondu* Region

Mamak is a huge *gecekondu* district, where 95% of the region is composed of *gecekondu* houses. According to the 1990 population census, the population of the central Mamak district was put forward as 401,584 while the population of towns and villages that were considered as part of Mamak was 10,600.[265] The region covers an area of 16,126 hectares and 6000 of it was a settlement area. Elmadağ, a small mountain for winter sports, shares a boundary with Mamak on the east; Altındağ, an important neighborhood of "old Ankara", on the north and northwest; and Çankaya, the middle and upper middle class district of today's Ankara, on the south and southwest.

A discussion of the development of *Mamak* as a huge *gecekondu* region where the Boğaziçi neighborhood is located, is important in order to understand the emergence of peripheral *gecekondu* settlements in Ankara. The region, today called Mamak, was a village of Ankara in 1884. After the establishment of the Republican Regime, Mamak village was acknowledged as a suburb of Ankara. The region is located in the eastern part of Ankara. This peripheral settlement was part of the Çankaya administrative district in the late 1940s and early 1950s and from 1950 onwards, there emerged *gecekondu* settlements in this region. In time, the population of Mamak village and other villages on the two sides of the Samsun Highway increased due to the migratory flow.[266]

The farthest village, on the axis of the Samsun Highway to the center, was Kayaş located in the western side of Ankara, Kayaş was a quite distant and isolated rural settlement in the 1950s. In addition there were the villages of Araplar and Üreğil. Mamak village was located closest to central districts such as Cebeci on the axis of Samsun Highway. Over time, these villages were transformed into large *gecekondu* neighborhoods and the distances between them were closed with the rapid expansion of each neighborhood. Parallel to Samsun Highway on the southern face of Mamak, small *gecekondu* settlements expanded and formed today's Tuzluçayır and Nato Highway regions that are quite close to Çankaya district, an upper-middle and upper class district of Ankara currently. The rural spaces extending from Kayaş village to the center of Mamak, in a three kilometer range; was occupied by *gecekondu*

---

265 Mamak Municipality (1992), *Belediyecilik ve Şehircilik Açısından Uygulama ve Projeleriyle Yeni Mamak: 1991-1992 Faaliyet Raporu* (The New Mamak with the Municipal and Urban Policies and Projects: 1991-1992 Action Report), p.5.

266 Ibid., p.5.

settlements from both directions towards the end of the 1960s.[267] It is difficult to discover which neighborhood developed first. Forerunner migrants selected certain lands close to the villages, taking into account various factors such as financial concerns, natural risks, transportation parameters and the suggestions of their acquaintances. In terms of landownership patterns, the region reflects a dual character. Some parts of it were state owned lands while other parts were vineyards, privately owned by the villagers. In this respect, the migrants either appropriated state owned lands or bought lands from the villagers in order to construct their houses.

Rural-to-urban migrants, arriving in the city by means of the Samsun Highway, preferred the Mamak region, located close to the route to their hometowns, as a suitable place to settle. Due to economic difficulties, particularly in the initial stages of the settlement, the physical proximity and the easily attained transportation facilities connecting Mamak and their hometowns constituted the major factor in the choice of settlement. As residents of Mamak, they were easily able to go to their hometowns, since the intercity buses operate along the Samsun Highway. As mentioned above, this was an important phenomenon in the early periods of the rural-to-urban migration when people still protected their close social and economic ties with their villages.

The migrants in Mamak, to a large extent, worked long distances from their settlement, because the region was not close to the commercial districts. In Jansen's plan, Mamak region was described as vineyards and gardens, and this status as a green belt was to be preserved, according to the premises of the plan. The planners left no place for commercial districts in the plan, and therefore, no large commercial districts emerged in the region, according to the Mamak Municipality activity report in 1992.[268] In the general land use map of Ankara of the 1960s[269] the distance between the commercial districts and Mamak was easily observable. Therefore, being close to cheap transportation services seemed to constitute an important concern for the migrants. Mamak was connected to Ulus, one of the two centers of Ankara at that time, the center of the Republican Regime at first and then the center of the fringes, by railway. This made transportation costs for *gecekondu* dwellers affordable. The need to travel to work and the emergent shopping needs were met by the railway at the time. Except for the *gecekondu* districts at the center, no other peripheral *gecekondu* district could utilize the railway in the 1960s and the 1970s. In this respect, the expansion of Mamak as a huge region does not seem to be a coincidence.

267 Akçura (1971:72).

268 Mamak Municipality (1992:42).

269 See Akçura (1971:75).

3.7 Municipal Experiences of Mamak Region

During the *DP* (Democrat Party, conservative party) era, only 3.5% of Mamak region had construction permits. As mentioned above, in the years where the influence of the social democratic tradition was felt by local government, there was greater attention paid to peripheral settlements. In order to increase the frequency of dialogue of these regions with the central districts, both physical construction projects and social projects were put on the agenda. The transportation facilities increased and infrastructural services were provided in peripheral settlements as a consequence of the self-help community projects by the municipality.

With the establishment of a two-tiered municipal system, five main settlement areas in Ankara gained their status as autonomous administrative districts, having their own district municipalities but connected to the Metropolitan Municipality, financially and administratively. Law number 2963 acknowledged Mamak as one of the administrative districts of Ankara; along with Yenimahalle, Çankaya, Altındağ and Keçiören.

After 1984, with the assembling of the Mamak Municipality, the effects of local policies in the social spatialization of the region increased. The first mayor of Mamak Municipality was from *ANAP*. The first municipal period was highly criticized in the activity report of the succeeding period of social democratic local government in Mamak. According to the activity report of 1992, where the mayor from a social democratic political party was in charge, the preceding period 1984-1989 was defined as one during which construction plans were properly prepared and applied.[270] The civil servants I interviewed in Mamak Municipality mentioned the fact that they had started their careers in 1989 in Mamak Municipality with the appointment of the social democratic party. In this respect, they were also critical about the *ANAP* period and the administration of Gazi Şahin who became the mayor of the Virtue Party, an Islamist party descendant of the conservative factions of the Welfare Party in 1999. Gazi Şahin remains the mayor of Mamak and has been a member of the ruling *AKP* since 2003.

The difficulties I encountered in the search for documents and planning practices for each municipal period, and the complaints of municipal officials from different parties about each others' political interventions in the region, reflect the nature and effect of favoritism in local politics. Each municipal period of the different political parties neglects, denies, criticizes and conceals implicitly or explicitly the experiences and actions of the preceding local government. Moreover, with every new administration, a significant number

270 Mamak Municipality (1992:42).

of municipal officials are removed from duty and new ones are appointed under the administration of the new local government. The underestimation of the policies of preceding local governments by the new local government, when combined with the estrangement of experienced personnel who are members of the opposing parties, lead to a loss of accumulated local knowledge about the region and social structure of *gecekondu* settlements. This fact, when combined with the difficulties in determining the complicated nature of landownership patterns in the region, brought about the improper implementation of construction laws and policies. This fact further triggered the emergence of unplanned and unintended spatial developments.

As mentioned in the previous chapter, the 1984 *gecekondu* law initiated an immense amount of legalization of *gecekondu* settlements. During the *ANAP* Period, the Mamak Municipality distributed "*tapu tahsis belgesi*" through the mediation of private technical offices under oath without making a prior in-depth investigation of the complicated land ownership patterns in the region. In a sense, the municipality applied the law without serious consideration of the local context. The tactical acts of some migrants who were economically and politically more powerful turned the legalization process to their advantage within the context of nepotistic policies.

The most ironic aspect of the research conducted in Mamak Municipality was to observe the hesitancy of those municipal officials who criticized the favoritism going on in Mamak for years in a very harsh manner on the one hand, but mentioned the incidents in which they led certain practices of favoritism to happen in past on the other. When they talked to each other following the end of our formal interview, they narrated how they had asked demolition officials to ignore places where their relatives were residing during those years.

The common objective of all municipal periods was to transform Mamak and integrate it as part of the city. Therefore, mayors gave utmost importace to the construction of modern recreational public facilities and the realization of urban renewal projects that aimed at solving all the complications of land and house ownership patterns in the neighborhoods. The objective was to systematically open them to modern construction.

In the early 1990s, under the administration of the mayor from the social democrat political party, the act of legalizing the *gecekondus* and distributing title deeds according to 1984 *gecekondu* law had been completed to a large extent. The places legalized for construction increased to 77% during this period and *gecekondu* dwellers were given title deeds. Private offices under oath defined or measured the amount of the land *gecekondu* dwellers had a right to own legally. This was done by taking the word of the *gecekondu* dwellers who determined the size of the plot on which the *gecekondu* house had been built. This practice was intrinsically open to exploitation, both on the part of the *gecekondu* dwellers, and by the architects and engineers taking the measurements, as mentioned in the previous chapter. The *gecekondu* dwellers, at times, may have misled officials on the status of the land on which they had built their houses. They may have declared the previous status of the land as owned by the state or municipality, although, some of these lands turned out to be privately owned. This caused a major problem for municipalities because people who were the real owners of these lands made their claims by carrying the issue to the court. This constituted the main impediment behind the planning activities of the municipalities to open these regions to formal construction. According to the reports of the municipal officials who took on active parts in the legalization process, they experienced real difficulty in the process of distributing title deeds. The declarations of the *gecekondu* inhabitants about the status of the land on which they had built their houses did not always coincide with the truth.

Urban renewal projects, particularly in the *Alevi* dominated neighborhoods, were initiated during the social democrat municipal period. The activity reports related to this period also give clues about favoritism in the form of giving services mainly to *Alevi* dominated neighborhoods. In some cases due to the *Alevi-Sünni* conflict in the region, municipal officials may have got a reaction from the *gecekondu* inhabitants as a result of the prejudices on the part of each party. Social democrat officials reported incidents where they were threatened with guns and were targeted as "unbelievers" during the implementation of a decision to destroy a mosque in a *Sünni* dominated neighborhood on land that is topographically unsuitable for construction. The demolition of *gecekondus* that were occupied by people, in opposition to the ideological stand of the mayor, led to a reaction on part of the migrants. Foremostly, such demolitions meant a loss of their homes and, besides, they took the actions of civil servants as an attack against them by their political and sectarian rivals.

When the activity reports of the religiously oriented mayor were examined, a concentration of municipal services in certain *Sünni* dominated neighborhoods were revealed. The urban renewal projects occuring in many neighborhoods constituted the main issue in the agenda of the Mamak Municipality. The activity reports reflect the main objective of these projects in the region as constituting healthy and sustainable living environments by solving the landownership problems of jointly owned lands.[271] Green areas and recreational spaces were supposed to be implemented within the framework of this new approach to urban renewal projects. Particularly in those parts very close to Çankaya district, urban renewal projects like Doğukent (East city) and Ege neighborhood were given utmost importance

271 Mamak Municipality (2001), *1999-2000 Yılı Faaliyet Raporu* (Year 1999-2000 Action Report), p.113,114.

because this reflected the opportunity for Mamak to change its image as a *gecekondu* district, and also symbolized the integration of the region with the city as mentioned in municipal documents. These two projects were located very close to the Mamak-Çankaya viaduct, reducing the physical and, hopefully, social distance between the two different regions of Ankara.

To sum up, the local municipal history is full of anecdotes about *Alevi-Sünni* conflict and favoritist policies that were operated by municipal governments. Favoritism seems to have constituted the main impediment behind effective planning in the region and triggered the uneven development. Therefore, some neighborhoods, as a result of their late integration into the legal estate market of the city, were deprived of the material and social benefits that came with systematic construction. The planning activity in Mamak region, like the macro planning activities in Ankara, most of the time followed the unintended developments rather than avoiding or preceding them. There were legal and institutional drawbacks to the planning activities besides the operation of local agents.

Favoritism in politics contributed to the spontaneity in peripheral settlements. Migrants, in some cases, tactically used favoritism to their benefit, while these acts may have worked to the disadvantage of other groups. Favoritist policies still dominate local politics. However, realizing important urban renewal projects in certain neighborhoods meant a lot to both inhabitants of the region and mayors, regardless of their sectarian and political affiliations, since it helped the integration of Mamak into the city even though Mamak, as huge *gecekondu* region, had transformed the public and political space of the city for some time.

3.8 Conclusion

The construction of Ankara as the capital city of the Young Republic can be accepted as constituting one of the few examples of the imposition of a radically modernist, nationalist social and physical space from top to down, with a serious neglect of the existing social context of the city. Ankara's urban history, which dates back to the Phrygians era, and the town's local people, their representational spaces and spatial practices were neglected to a large extent during the imposition of this holistic modernist project.

The motivations behind the need to create such a western city reflects traces of the uneasy and hesitant relation of the Republican Regime with Western European cities. As stated by Sibel Bozdoğan[272] "Turkish architectural

---

272 Bozdoğan, S. (1997), "The Predicament of Modernism in Turkish Architectural Culture" in Bozdoğan, S. and Kasaba, R. (eds.), *Rethinking the Modern Project in Turkey*, Seattle: Washington University Press, p.137.

culture of the 1930s adopted the formal and scientific precepts of Western modernism and yet posited itself as an anti-imperialist, anti-Orientalist, and anticolonialist expressions of independence, identity and subjecthood." That is the reason why the construction of Ankara as a representation of the modern Turkish state can be accepted as an act *for* and *against* the Western world.

The attempt of the Republican Regime to build a model city reflecting modern urban space and lifestyle, and the continuing effect of such a radically modernist ideal created a major rupture between the "lived" and "conceived" space of the city. As mentioned in the introductory chapter, on the word of Lefebvre, the three aspects of his spatial triad- perceived, conceived and lived- go together and there are no antagonisms, oppositions and contrasts between these three aspects of the triad, rather, there are echoes, repercussions and mirror effects. This model of Lefebvre appears to be valid within the context of his references to mainly Western capitalist cities. However, in planned cities of developing nations, like Ankara, Islamabad, Canberra and Brasilia, there appear to be antagonisms, oppositions and contrasts between the three aspects of the triad. This is the underlying reason behind the intrinsic impediments and failures of such radically modernist planning experiences. In addition to the inherent limitations of a modernization strategy, the second reason for the failure is the financial and technical incapability of the developing states. The planning and construction of Ankara well reflects these impediments of the strategical realm.

Within such a context, the representational spaces or the lived spatial practices of the "weak" may attain capabilities not only to divert strategical space but, in some cases, to affect the course of planning activity, which was the case in Ankara. The emergence of *gecekondu* spaces dependent on the limitations of the strategical planning policies, over time seems to have altered the course of post-planning planning practices. The transportation routes and the changing functions of the two centers of planned Ankara can be given as examples of the capabilities of tactical acts in creating spatial practices that bring about contested spaces in the city.

After the early 1980s the spontaneity embedded in the appropriated nature of *gecekondu* settlements further strengthened with the emergence of favoritism in local politics. This will be dealt with in detail in the following chapters referring to the experiences and insiders' views in the Boğaziçi neighborhood of Ankara.

CHAPTER 4 - Initial Phases of Social Spatialization: Natural Space

4.1 Introduction

This chapter attempts to evaluate the initial phases of social spatialization in the designated research setting. The transformation of the research site from the vineyards and farms of Üreğil villagers, a mainly natural space, to a *gecekondu* region will be described. This narration includes the incidents pointing out the interconnected histories of the development of *gecekondu* space and Ankara. In addition, the spontaneous quality of the spatial development in the neighborhood will be undertaken by referring to the spatial practices and *tactics* that *gecekondu* people employ from within the system.

As mentioned in chapter one, critical evaluation of De Certeau's concepts of tactic and strategy via Lefebvre's critique of modernist production of space would constitute the theoretical framework of this analysis. Both theories, as critical of everyday life in modernity, appreciate the potential of "tactics" (De Certeau) or "users" (Lefebvre) in their capacity to create differential spaces within the system. However, in both this capacity is portrayed in a highly limited manner with an overemphasis on unilateral power relations between strategical and tactical realms, attributing a limited agency to the tactical realm. Both refer to squatter type settlements and their ability to create a differential spatial trajectory within modernist space of strategy; however, they mainly regard these spontaneous spaces as receptive or submissive.

As will be made explicit by the empirical findings of the research, the tactical acts of migrants from within the system via the everyday social spatialization of the locality seem to have a significant impact on the realm of strategical planning and have a transformative power, though limited, on city space. I will elaborate on the narratives involving the years 1950s and 1960s, the initial settlement years in the neighborhood, through the perspective of such a theoretical agenda.

I will deal with the research data related to this period on two different levels. First of all, I attempt to pursue the history of spatialization in the region depending on the totality of individual narratives and life stories. Since this study has a local historical dimension, the narratives of the respondents function as the main source of information in understanding the nature of social spatialization in the neighborhood.

At the second level, in combination with the theoretical framework mentioned above, I try to exemplify how migrants define their acts of survival, with regard to both natural factors and the supervision of security forces in the initial years of settlement and in the process of constructing their illegal shelters. These acts are described within the context of their positioning with regard to the realm of strategy and to the actions of other groups in the initial years of social spatialization.

To accomplish this task, I will first of all assess how different groups of migrants at specific periods employ certain tactics with regard to the strategical realm on the vertical level. The assessment will expose the weaknesses of the modernist planning attempts and policies from the viewpoints of the *gecekondulus*.

Secondly, with reference to the acts of *gecekondulu* on the vertical level, I will consider how power relations between different groups in the neighborhood operated by means of employing tactics vis-à-vis each other on the horizontal level. Some groups, depending on the context and time, may find chances to carry out "*near-strategical acts*"[273] in social spatialization of the locality with regard to the less powerful groups in the neighborhood. The less powerful groups[274] depending on the context, seem to act within this near-strategical space of the more powerful groups. Similar to the vertical model of strategy and tactic, the tactical acts of the less powerful groups within *gecekondu* society may bring unintended consequences for the more powerful groups that may change power relations in favor of less powerful groups in the coming decade.

The spontaneity of these settlements seems to emerge from these power relations at the horizontal and vertical levels. To signify the spontaneity embedded in the social spatialization of these settlements is not for the sake of exaggerating the power of the weak but to point out the loopholes of strategy particularly in creating a modernist space as was the case in Ankara. The most important impediment of the modernist strategy specific to Ankara was to undermine the social context of the city while planning and implementing a hypermodern space from top to down. The strategy excluded certain groups from the beginning by creating an urban space

273 I use the term "near-strategical acts" referring to tactics of more powerful groups vis-à-vis the less powerful ones in the neighborhood. They have the ability to make future projections and shape the (social) space of the research setting to an extent. This ability on part of the powerful groups necessitates a concept that transcends the concept of tactic as defined by De Certeau.

274 I rely on a dynamic understanding of power. In accordance with this, certain *hemşehri* and sectarian groups have lost and gained power depending on the political and economic context of social spatialization in the neighborhood over time. So the less powerful groups at a certain moment in the history of the neighborhood may turn into the relatively powerful groups in the coming period.

suitable only for the middle class western urban life. This aspect of strategic planning when combined with other impediments of urban policy and planning as discussed in chapter three seem to provide a more extended space for the tactics of *gecekondulu*. *Gecekondu* construction was an unlawful process at first; however, this informality or unlawfulness cannot be defined simply as being outside and marginal to the formal procedures and institutions of planning and urbanization policies.

When the urbanization history in Ankara and in other cities of Turkey is taken into consideration, it becomes clear that, to a great degree, planning strategy responds to spontaneous developments. As İlhan Tekeli states, "the ex-post regularization of unplanned, de facto developments" is one of most important planning activities that constitutes the planning cycle in Turkey.[275] Therefore it seems impossible to think of strategy separate from the tactic of the "other". The early history of Boğaziçi will be described within such a theoretical framework.

4.2 Survival Tactics within the Space of Strategic Loopholes: Urban Agriculture, Railway and Highway

As mentioned in the previous chapter, Ankara was a planned capital and was constructed mainly as the center of state bureaucracy. Outside the "urban center" that was built and planned by the *Kemalist* Regime, the spaces in the periphery of the city exhibited mostly "rural" qualities. Although an immense planning and construction activity in Ankara had taken place mainly by the design of urban middle class residential and bureaucratic districts, a large amount of land in the periphery of the city was left outside the context of planning practically as unorganized and unmanageable natural spaces in the 1950s and 1960s. This seems to be one of the very first omissions of planning in dealing with the issue of *gecekondu* during those years.

This area consisted of farms and vineyards, property of small villages on the periphery of the city, in combination with state owned lands. The vineyards and farms were mainly concentrated nearby the branches of Hatip River from Cebeci neighborhood to Kayaş village, where the vineyards of Üreğil villagers were also to be found. The vineyards of Üreğil villagers would later constitute the lands of Boğaziçi *gecekondu* neighborhood. In general, the periphery of Ankara city belonged to villages and therefore it was on lands on which agricultural activities were still being practiced

275 Quoted in Keskinok, Çağatay (1997), *State and the (Re)production of Urban Space*, Ankara: Metu Faculty of Architecture Press, p.86.

until the 1960s. When compared to İstanbul, the development of industrial and finance sector was negligible in Ankara in spite of state initiatives in founding an industrial infrastructure. The periphery of Ankara at the time was included in the context of the plan as the green belt of the city. Even after the planning activity had been initiated, the rural activity in the periphery of the city still constituted a pull factor for the migrants coming from the villages in closer towns and cities as seasonal labors.

The document representing the municipal boundaries of Ankara in 1926 cited the vineyards and farms of Üreğil villagers as *Davulcu Bağları* (Davulcu Vineyards) as being within the limits of the city.[276] In the 1930s there were two important villages other than Üreğil in the region, namely Kayaş and Araplar. These three villages have grown into large neighborhoods which are still associated with their original names today, as mentioned above. Üreğil villagers owned big plots of land and were engaged in large-scale agricultural activities that needed a substantial labor force in the early Republican Period. As the 95 year old son of the last landlord describes, there were four or five seasonal rural laborers staying in every house of the village in the most demanding periods of the agricultural calendar.[277] Üreğil vvillagers produced tobacco for *Tekel,* the state firm that was the only producer of tobacco products and alcoholic beverages at the time. Additionally, villagers grew variety of vegetables and sold them to the military headquarters nearby and also at vegetable bazaar in the old city center.

As mentioned in the second chapter, many of the classical *gecekondu* studies supposed a transitory model in migrants' adaptation to city. These studies assess the migrant's involvement with rural activities in the early settlement period in the city as either a resistance to giving up their rural ways of life, or an adaptive/supporting economic activity for subsistence. Certainly these findings have some validity, which varies according to each individual *gecekondu* district. However, the course of events might follow a different path as in the case of the formation of the location chosen for this research setting. Contrary to the conventional studies, the large-scale agricultural activities in the peripheral villages of Ankara directly brought about the first encounters of the migrants with an "urban" context. The underlying motivation behind migration decision of the first wave of migrants depended on the supply of rural activities on the edge of Ankara.

---

276 Şenyapılı, Tansı (2004), *"Baraka"dan Gecekonduya: Ankara'da Kentsel Mekanın Dönüşümü 1923-1960* (From Shed to *Gecekondu*: The Transformation of Urban Space in Ankara), İstanbul: İletişim Yayınları, p.307.

277 Extract from field research note no.44, 17 May 2004.

This process contradicts with the modernization approach in general which delineates agriculture and urbanization as conflicting issues.

The first residents of Boğaziçi neighborhood had been the temporary farm laborers who worked for Üreğil villagers. The migrants constituting the majority of the neighborhood today, people from Cemele village in Kırşehir province, worked once as laborers in the farms of the Üreğil villagers and they were the ones who laid the foundation of the setting. The vineyards of the Üreğil villagers in combination with state owned lands had been appropriated by the migrants from Kırşehir and nearby villages of Ankara in the first wave and later by migrants from Kırıkkale, Çorum, Yozgat, Çankırı, Sivas and Tunceli.

The journeys of migrants from Cemele / Kırşehir to Üreğil lasted four or five days in those days, therefore laborers from Cemele, most of the time stayed in houses of Üreğil villagers. This intimate relation between Üreğil villagers and migrants from Cemele explains villagers' decision to sell small plots of land from their vineyards to migrants in the subsequent years. The first wave Cemele migrants began to occupy the region by building their *gecekondu* houses, and thus constituting a small community in these early years. The first *gecekondus* in the neighborhood were those built on the vineyards or farms of the "native" Üreğil villagers.

As was described in the introductory chapter, a long street now divides Boğaziçi neighborhood -located in a small valley and was once a river- where the market place is now situated. The vineyards of Üreğil villagers were located on either side of the valley in the early settlement years and that is the reason why today migrants from Kırşehir mainly reside in there. As understood from the narratives of the first wave migrants, Üreğil villagers were more willing to offer their lands from their vineyards rather than from farms due to the lower agricultural economic value of vineyards. Thus they could be sold relatively easily to the seasonal laborers, and the sale was also a show of respect towards the migrants from the villagers.

Over time, this process gained impetus; the relatives of the migrants continued to migrate and buy lands from villagers or, in some cases, built their houses on state owned lands adjoining the vineyards and farms. The migrants employed tactics as in the form of "colonizing space" initially during this period; this was the first attempt of migrants to establish themselves in the city. The main motivation behind these migrants was not to search for an "urban" job but rather to work in the rural type of employment in the natural space / environment of this research setting. In contrast, even the empirical studies that consider the interconnectedness of the formation of *gecekondus* with economic and political context of the city suppose a certain discontinuity between urban and rural contexts. In explaining the history of *gecekondu* formation in Ankara, Şenyapılı[278] defines the early migrants from the villages close to Ankara as searching for seasonal "urban" work in addition to their rural incomes from the villages as the main actors of *gecekondu* formation in the topographically unsuitable settlement locations such as Mamak. The temporary visit of rural migrants in search for work in Ankara was only linked to the push factors in the village by Şenyapılı, who gave no account of these temporary workers employed in the rural type of facilities in the periphery of the "planned city".

The topographically unsuitable places for settlement constituted a suitable natural habitat for urban agriculture and animal husbandry in Mamak *gecekondu* region in the 1940s and 1950s. Even the contemporary literature on shanty towns in Latin America and other third world cities confirm that urban agriculture at the edge of cities on lands and water areas in the urbanized sphere that are available for agricultural use of surfaces constitute the main source of income or subsistence for poor people.[279] Similar to this recent finding, urban agriculture and animal husbandry in the periphery of Mamak constituted the main means of survival in the most economically depressed period of the neighborhood history, as defined by migrants. The misery of those early years was related to poverty coupled with the lack of certain services in the neighborhood. The basic needs of survival, such as dwellings, nourishment and heating constituted the main concerns in a natural environment. Due to these early deprivations, the relations with nature in the initial periods of settlement involve both a struggle with the natural elements of it and at the same time the need for utilization of its resources to the fullest extent. These survival tactics with regard to nature made use of the space of strategy that had been left as unorganized and uncontrolled by the planning practice itself even after the settlement of migrants there.

Migrants variously engaged in small or medium-scale animal husbandry by utilizing the natural environment, working as wage agricultural laborers in the farms of the villagers, and growing vegetables and fruit in their gardens for subsistence. An anecdote concerning the utilization of natural resources around by an elderly woman who migrated from a village close to Ankara to the region is given below. This anecdote is also important in understanding the poverty experienced by many families particularly in the early years of settlement:

> May God keep us from returning to these days. We experienced great hardship at the time. On the other hand, the taste of yogurt

278 See Şenyapılı (1985:114); (1998:305) and (2004:73).

279 See Madaleno, Isabel (2000), "Urban Agriculture in Belém, Brazil," *Cities*, 17(1), p.73.

> and milk was different and everything was fresh then. When we first come here, there were vineyards and orchards everywhere. Your uncle [referring to her husband] sold butter at the time. We heated the house with *tezek* (dried cow dung). We burned them in the stove. We cut grass and thresh in the farms here, work as laborers. My dear. Life was hard back then. We lived a life of misery at the time.[280]

Particularly in the narratives and memories of women, the early years of settlement take on a special significance. In the natural flow of the structured or unstructured interviews with older women, they often refer to these early years in a highly emotional manner. This seems to have a relationship with the fact that women had an extremely significant role and overload in household's economy during those years. In fact, survival by appropriating[281] the means of natural space in Lefebvrian sense constitutes the main concern of migrants' daily lives in "urban" context. The exploitation of natural resources includes tactical decisions and behavior that was forced on them by the limited possibilities and danger provided by natural space. The domestic labor that exhibits such tactical acts in relation to the nature had an extended role in everyday life.

Children as well as women were active in agricultural and stockbreeding activities, and putting animals out to pasture was an important activity conducted by the children. The following anecdote is a typical description of this existence with nature and animal breeding given in response to questions about the general physical conditions of the neighborhood, in the 1950s especially by respondents over 50 years old.

> We bought cows, and went to put animals out to pasture. We went to the hills opposite. We took them to the municipal garbage

280 Extract from field research note no. 11, 4 March 2003 (Interview with an old lady from the village of Ankara. Her son has a large store selling household appliances and owns a large apartment house in the neighborhood).

281 Lefebvre supposes a distinction between domination and appropriation of space. Domination by technology that is affiliated with capitalism's strategical abstract space tends towards non-appropriation, destruction. See Lefebvre (1998:343). Appropriated space is a natural space in order to serve the needs and possibilities of a group. See ibid., p.166. Rather than dominating the space by employing future projected strategies, migrants appropriate the resources of natural space as an immediate tactic for survival considering the deprivations of initial settlement period.

> dump. That is to say, we began to do stockbreeding. We sold their milk. Since here it was all green at the time, feeding them cost little. We were so poor then. When we went to school with my brother, the flies set down on us since we smelled like milk.[282]

To sum up, the available natural space in the periphery of the city and the ongoing rural practices in this periphery constituted a habitat and a means of survival for migrants. As mentioned in the second chapter, due to the planning priorities and conditions of the city in the 1940's, industrial development in Ankara was highly limited, providing job opportunities mainly in three sectors of construction, trading and service.[283] In the following decades, low profile service jobs in the state sector and in trading constituted a remarkable part of *gecekondu* society's occupational composition, parallel to their experiential maturity in the city. As gathered from the research data, economic dealings resulting from animal husbandry and agriculture in *gecekondu* space seemed to last longer than conventionally defined, and was seen as a means to accumulate capital for other jobs, particularly in the trading sector. The early experiences of the migrants in the "rural" environment of the periphery of the city not only raise certain questions about the supposed dichotomy between rural and urban spheres, but also about conventionally defined ways for *gecekondulu* in integrating with the job market of the city. Agricultural activity in the periphery of Ankara constituted the primary step in their integration with the city economy.

In addition to the availability of natural space in the periphery of Ankara, the development of Mamak as a large *gecekondu* region had two reasons for the strong spatial references on a macro scale. As mentioned in the previous chapter, the first one is the presence of the intercity train service in Mamak that sustained cheap transportation to the city center from the periphery from the 1930s onwards. The migrants often mentioned the presence of the railway as constituting one of the main motivations behind their settlement decisions.

The second reason is the presence of the Samsun Highway dividing the Mamak region into two major sections and its status as the main highway connecting the region to neighboring Central Anatolian cities from which most of the settlers of Mamak migrated. As Mike Davis[284] precisely suggests, housing choice is a result of trade-offs for the migrants everywhere in the Third

282 Extract from field research note no. 7, 16 Oct. 2002 (Interview with the owner of a *bakkal*).

283 See Şenyapılı (2004:73,170).

284 Davis, Mike (2006), Planet of Slums, New York: Verso, p.27.

World. "For some people, a location near a job, - say, in a produce market or train station, - is even more important than a roof".[285] That spatial proximity to a cheap means of transportation in two directions, both to the city center and to their villages or their hometowns constitutes the main motivation behind migrants' settlement decisions is an often-mentioned fact during the course of the research. The Mamak region is the most suitable place to settle for migrants in order for them to visit their villages of origin as easily as possible. Some respondents mentioned this spatial proximity by making jokes about these initial choices of where to settle. When I asked a middle-aged man about his ideas of settling in Boğaziçi, he answered in the following manner:

> I often get angry with my father. As if there was nowhere other than Boğaziçi, they had come and settled here. But my father was unemployed at the time and they should have thought about whether living in a *gecekondu* settlement would be easier and suitable. I often teased my father by saying: "On the way to Ankara from Kayaş, you saw the city sign and immediately settled here, I think." It was a *gecekondu* settlement at the time and everywhere was vineyards and orchards. Here was the river. There was no road at the time. My father probably was thinking that he could work in the farms here. If he had settled in a different neighborhood he might have lived in real poverty.[286]

Nevertheless, the construction of the intercity railway in the late 1930s and highway in the 1950s mainly emanated from the objectives of a modernist strategic planning. As mentioned in the previous chapter, the development and improvement of the railway was one of the main strategies of the Republican Regime in their attempt to open the introverted and conservative structure of Ankara as a central Anatolian city to the modern world, in social and economic terms.[287] Towards the end of the 1930s the train station was opened in Cebeci and the railway route between Mamak and Kayaş was started, mainly to carry the middle class families of the new Republican capital to picnic areas in Kayaş, was one of the main natural recreational spaces in those days. Unintended as part of the planning objectives of the subway, the initial tactical decisions of the migrants in the city brought about the emergence of *gecekondu* settlements as spontaneous spaces in that particular region.

---

285 Ibid., p.27

286 Extract from field research note no. 28, 17 June 2003 (Interview with a shop owner from Yozgat).

287 Şenol-Cantek (2003:42).

In a manner similar to that of the railway construction, the Samsun Highway was constructed as part of this modernist-planning attempt. As mentioned in the second chapter, Democrat Party (*DP*) had a hesitant approach to low-income settlements and urbanization policies in general. In spite of their early populist policies and their condoning *gecekondu* construction in the periphery of the city, their urban policies in the late 1950s included radical and despotic modernist interventions in urban space. The construction of big boulevards, highways and modernist buildings, and the demolitions in the "cancerous" neighborhoods to quote the late Menderes, were the main activities of the *DP*'s urban political agenda in those years. The construction of the Samsun Highway was part of this radical modernist agenda of the *DP*. However, over time, the construction of the Samsun Highway also brought unintended consequences for the agents of strategical planning, particularly by altering the real estate values in the region.

In the 1940s the Boğaziçi neighborhood, in close connection with the ownership patterns of the vineyards, developed from the areas closer to the main road. The natives of Üreğil village owned all the lands on both sides of the Samsun Highway from the Boğaziçi neighborhood to Kayaş, approximately two kilometers in length. The merging of small and scattered migrant communities on those lands into Boğaziçi and other neighborhoods took impetus, beginning in 1952 with the construction of the highway. The value increase in real estate, with the construction of the nearby highway, brought about a declining importance of agriculture in the region. Consequently, the villagers on the periphery sold their lands to the migrants coming in large numbers from different parts of Central Anatolia and from other *gecekondu* neighborhoods of Ankara. In a context where agriculture in the periphery of Ankara had begun to lose its economic significance compared to previous years, *gecekondu* construction followed a more rapid course, beginning in the late 1950s. Besides, the first *gecekondu* law that particularly gave reference to the unlawful settlements of Ankara was passed in 1948. By means of the law, these unlawful settlements called as Mamak-Belkeriz *gecekondu* region that were diffused over a 650-hectare area were taken within the context of amnesty.[288] This also had a triggering effect on the development of the Boğaziçi neighborhood in the 1950s within the context of this vertical, dialectical and interconnected relationship between urban policies on the strategical level and the acts of migrants on the tactical level.

---

288 See Şenyapılı (2004:133,135).

4.3 *Gecekondu* Construction: Tactic versus Strategy on the Vertical Level

As a consequence of the intensity of *gecekondu* construction during this period, the *DP* government took more severe legal measures in the late 1950s in contradiction with its previous populist approach towards *gecekondulu*. In this period, the demolition incidents in Ankara increased in number. Menderes made declarations that the city had been long left uncontrolled and the neighborhoods that he thought of as the sources of crime should be tamed.[289] These severe measures and the similar ones following it in the subsequent years could not prevent the accelerated growth of *gecekondu* construction as mentioned in chapter two. The narratives of the respondents about the construction of their *gecekondus* provide some insights about the loopholes in the legal application of the laws and the tactics of *gecekondulu* that utilized these loopholes; this made the growth of *gecekondu* spaces inevitable.

As understood from the accounts of the migrants, the relationship between them and those state officials who had the task of demolition seems to have been an unwritten game with agreed rules and mutual understanding at an informal level. Each party performed in order to avoid tension on part of the other. In this way, migrants avoided any punishment or violent act. The relationship between the state officials responsible for the supervision of unlawful constructions and migrants was, in general, unexpectedly "loose" or "unmanageable". Migrants mentioned the informal nature of a relationship that was mainly a consequence of the financial interests of officials. Officials often took bribes from migrants in order to overlook illegal activity. The second reason for this informal relation was the emotional bond between the migrants and certain officials who were also *gecekondulu* and shared the same experience of surviving in the city.

In some cases, where the relationship between parties took on a problematic nature, the incompatibility did not usually emerge from the proper applications of the rules and procedures by the officials but from the tightening of this loose relationship as a result of some personal issue between the migrant and official totally outside of its legal context. Depending on the narratives of the migrants, this tightening may have come from different sources, like sectarian or hometown differences between parties, incorrect actions on the part of each party in violation of the game-like relationship or because of a personal dislike. There was only one unfortunate event encountered in the narratives of the respondents where a *gecekondu* dweller murdered an official after his fourth attempt at demolition. This incident

---

289 See Şenol-Cantek (2006:51).

was related to the migrant's refusal to give a bribe to the official and the public announcement of the official's requests for bribes. In this event, the rules of the game-like interaction between the *gecekondulu* and the official were totally broken. Despite the sensitiveness of the issue, giving and taking bribe was made totally explicit in the stories of the respondents. The demolition incidents reached their peak in 1961, just after the military intervention, during the temporary government of Cemal Gürsel. One old woman mentioned the importance of giving bribes to protect one's *gecekondu* during this specific period in the following manner:

> The military intervention had just been taken place when we came here. Soldiers used to come here, we served tea, and they said nothing about our *gecekondu*. Then, one day while I was carrying a very heavy wool bed on my back, I saw the demolishers around our house. I was scared; I remained motionless for a very long time with the bed on my back. Now, and back then the demolishers had only one concern. When you gave money, nobody demolished your house. When they run out of money, they came for the claim of destruction.[290]

In sum, depending on the stories of the *gecekondulu* about demolition, under normal conditions, each party acted in compliance with the invisible and silent rules of the game. The most often voiced tactic of the officials was to destroy some part of the *gecekondu* that did not lead to the total destruction of the house. In doing so, they seemingly accomplished the duty of demolition, but they caused no real harm to the *gecekondu* dweller since the damage could be repaired in a couple of hours.

Consequently, *gecekondu* construction and its prevention from demolition can be read as a showcase of *gecekondulu*'s tactics for survival within the system. In that sense, it seems convenient, in this part of the chapter, to give a more in-depth account of the tactics *gecekondulu* employed while constructing their houses and the consequences of these tactics. As noted from the narratives of the respondents, the first one of these tactics was to construct *gecekondus* in a very limited time span so that they could avoid the "selective" attention of officials whose duty was to apply the demolition decisions. The attention of the officials can be defined as "selective" because, as understood from the construction stories, demolition priority was given to the houses that had been most recently built or that had not yet been completed. The criterion of

290 Extract from field research note no. 36, 20 July 2003 (Interview with an old lady from Kırşehir / Cemele).

living in a *gecekondu* for a long time is narrated as one of the most important reasons for the migrants to resist the demolition decision. In this sense, the *gecekondus* that were built most recently carried a greater risk of demolition when compared to others. The narratives of municipal officials point out a practice, which they often encountered, during demolition incidents. Just after constructing their *gecekondus*, migrants planted their gardens with already flowering plants borrowed from their neighbors in order to give the impression that they had stayed at the house for months.[291]

The second tactic of migrants, in relation to the second condition of being "selective" on the part of the officials, was to build houses as small as possible, mostly in the shape of a "coal cellar" rather than a house. The initial construction of the house in the form of a "coal cellar", as commonly named by *gecekondulus*, may have served as a legitimate reason for officials to ignore the house. Almost all of the respondents described a resemblance between the initial shapes of their house with a coal cellar. "We built the house as a coal-cellar at first" is an often-mentioned sentence at the beginning of the construction stories. The resemblance of the house to a coal cellar would have constituted an excuse for them in the game-like relationship they formed with the official as mentioned above. The late-comer migrants, most of the time, constructed their houses near their *hemşehri*'s already established *gecekondus* so that they could present the small house as a coal cellar of the big *gecekondu* to the officials. In this sense, the late coming migrants were obliged to use the near-strategical spaces of early comer migrants as a survival tactic, in order to escape the demolition incidents. Unless the family was wealthy enough to build a two or three room *gecekondu* in one night, the *gecekondu* houses were constructed as small one-room building at first. *Gecekondus* took their final form over time after the addition of rooms convenient for the needs of the family. Similar to the development of these neighborhoods in an unplanned and spontaneous fashion, the construction of each *gecekondu* house totally depended on the experience of building, destroying and rebuilding. The outlook of the *gecekondu* reflects the effects of this experimental and unplanned construction process.

The third tactic, and perhaps the most important one for the migrants, was to choose the most suitable place to build their *gecekondus*, not only to escape from the attention of the officials but also to sustain the best possible living standards. The third decision can only be defined as a tactic if we approach it from the perspective of relations between state officials and *gecekondulu*. However, if we consider the power relations within the community, the decisions about settlement contained near-strategic acts based on future calculations for

291 Extracted from field research note no. 43, 15 Jan.2004.

more powerful groups, like the early-comer migrants or the migrants from hometowns who were densely represented in that neighborhood. I would like to give an extended account of the first and third tactics below.

4.3.1 "GECE KONDU" Built Overnight by the Community: Hemşehri Relations Employed as Tactics Operating on the Vertical and Horizontal Level

As mentioned above, houses under construction had a higher risk of being torn down when compared to completed ones. *Gecekondu* law number 486 gave full authorization to state officials for demolition of *gecekondus* that were under construction. For the completed ones, though, officials had to have a court decision, which was a difficult, and long process. Mostly it resulted in the protection of the house.[292] Therefore, under normal circumstances the demolition priority was mainly given to those houses under construction and it was essential to complete the construction invisibly. This invisibility could only be sustained at nights and through rapid construction. *Gecekondu* construction depended on the good management of time. The concept of "gece kondu" itself reflects its ontological status of existence that depends predominantly on time rather than space. In this sense, the early conditions of its existence very much coincides with the concept of tactic as defined by De Certeau. In this definition, the weak or the "other" of the modernist discourse could only find some place in the space of strategy by consuming it or, in Lefebvrian terms, by appropriating it in a differential manner by means of the good management of time. Therefore, as defined by the migrants, there was an immediate necessity to move into the house after the construction so that the officials would have the impression that the migrants had long been living in the house. As told by an old couple, moving into the house immediately seemed to be a necessity for the migrants at the time:

> When we first come here, we cut trees, bought briquette and sand. We gathered all our friends and acquaintances and worked collectively. Then the demolishers came. To prevent the demolition, we put haircloth and oar on the house under construction. We needed to move in immediately but we were afraid; the construction was new and everywhere was wet. We couldn't go in; we thought it was going to fall down. My brother stayed there the first night, in order to prevent the demolition.[293]

---

292 See Şenyapılı (2004:191).

293 Extract from field research note no. 31, 26 June 2003 (Interview with an old couple from Yozgat).

The rapid construction could only be sustained with the unconditional help gathered from the community. Depending on the accounts of the migrants, if no attempt of demolition took place, ten or more people were needed to complete a *gecekondu* with two rooms in one night. When I asked the migrants to tell of their experiences of *gecekondu* construction, they mostly referred to the miserable conditions and deprivations at the time. Most of the migrants defined this period as their most desperate and poor period. The feelings of poverty not only emerged from the fact that it was very difficult to attain water, construction materials or tools at the time, but as the narratives indicate, it was also difficult to get help in terms of human labor. The well-to-do migrants said that they had come to the region with a certain amount of money, and these migrants mostly mentioned their hiring workers during *gecekondu* construction, alongside the help given by close relatives. However, poor migrants were in direct need of all the material and non-material help they could get. It was at this point, while they were talking about these needs, that the definitions of close community were made, with reference to the nature of the help, if there was any. I initially was aware of the effect of power relations and competition for scarce resources between different groups at this point of the conversation, though they rarely gave any references to power relations and competition explicitly. They, rather, used a wording that associated their feelings of disappointment, helplessness, loneliness, and shame. An old woman from a small town of Ankara lacking a community of villagers or relatives describes their feeling of loneliness in the late 1950s as follows:

> When we first settled here I never went out of the house. Nobody helped me at the time. There were people mostly from Kırşehir. Anyway there were very few people around here at the time. They never helped. I still do not visit them and they do not visit me. I only got a water cauldron from a neighbor once. That was all.[294]

The first wave of migrants appropriated the resources of natural space and constructed a working system of survival for themselves and their community. As mentioned above, migrants from Cemele village of Kırşehir were the founders of the neighborhood. They were powerful because of their early appropriation of resources and the unconditional mutual support provided by the village community. Therefore, because of their experiences in their neighborhood and the city of Ankara, the behavior of these relatively powerful groups involved near-strategical rather than tactical desicions. The

294 Extract from field research note no. 11, 4 March 2003.

reason for labelling these acts as near-strategical is that they appeared to appropriate and control space with some future projections and calculations. They began to control the social spatialization in the neighborhood to the benefit of their *hemşehris*, which actually goes beyond the definition of tactic in De Certeauan terms.

Early, well-to-do migrants excavated wells in their gardens. As can be understood from the narratives of the respondents, most of the time, they want to reserve this source only for their close relatives rather than neighbors. Permission to use these wells seemed to constitute the main source of conflict in public space during those years. In most cases, the house owner gave permission only to relatives or people from their villages of origin. As a consequence, many, mainly female neighbors came into conflict over the issue of water. A shop owner from Yozgat gave an example referring to his mother's experience:

> When we settled here, naturally, we had no water. My mother often recalled an incident when she went to a house of our neighbor behind and took water from their well. According to my mother, the owner of the house took the water bucket from her hand and poured it on the ground with anger. I remember how upset she became on the occasions she remembered the incident. The reason for that is we are from Yozgat. 90% of the population here is from Kırşehir. People from Kırşehir resented people from Yozgat or Kırıkkale having settled here.[295]

As understood from the stories of the respondents, even the migrants from Kırşehir were reluctant to share among themselves their resources and labor if they are not from the same village or close community. However, when the nature of solidarity among *Kırşehirlis* is examined through their stories, they rarely voice any systematic denial of willingness to share resources with their neighbors from other hometowns. However, this emerges as a fact from their narratives that there had been solidarity only among close kin in the 1950s and early 1960s, a time of scarce resources.

The definition of being *hemşehri* literally refers to being from the same hometown or in other words to the migrants coming from the same province such as from Kırşehir, Kırıkkale, Çankırı, Çorum, Yozgat and Sivas. The definitions of *hemşehri* in relation to the representations of each province in the narratives of neighborhood inhabitants and stereotypes of these provinces are discussed in detail in the first chapter of the book. During the

295 Extract from field research note no. 28, 17 June 2003.

initial period of settlement, unlike its wider conventional meaning, *hemşehri* as a concept refers in the narratives of the migrants to a more restricted group: extended family, relatives or to people from the same village, which is a much more limited definition of the term.

This becomes apparent with the stories of people who had migrated from the densely represented provinces in the region such as Kırşehir and Kırıkkale, but had been excluded from the close community network due their being from another town or village of the same province. This limited definition of *hemşehri* led to feelings of isolation, expressed by the wife of the first *bakkal*, a well known member of the community:

> There were people from Kırşehir. But nobody helped us [during the construction of their houses and *bakkal*]. Though we earned less at the time, we always had money, although we suffered a lot we were never in need of anybody, thank god... There are a lot of people from Kırşehir in the neighborhood, but they are from a different town. We're from Kaman [another town of Kırşehir]. People from Kaman are not considered as *Kırşehirli* according to people from Cemele.[296]

Unconditional mutual material and nonmaterial help and solidarity are frequently encountered in the narratives of people who migrated from the same and densely represented villages. As mentioned above, this help was unconditional, involving sharing of all resources and exploiting every opportunity for mutual benefit. The significance of the nature of solidarity among migrants of close kin and relatives can most clearly be seen in *gecekondu* construction, due to the necessity of rapid construction. The difficulties of attaining water for the sake of constructing *gecekondu* and domestic tasks are the most often evoked memory in the narratives of the older women, and women especially refer to the solidarity among close kin or migrants from the same village. It is women who mostly use the restricted definitions of *hemşehri*, since they generally have a more restricted social environment when compared to men. As concluded by some research findings, due to restricted living environment, women consider relationships with neighbors even more important than kinship ties when compared to men.[297] They acknowledge people as their *hemşehris* only when

296 Extract from field research note no. 33, 30 June 2003.

297 See Gökçe, Birsen, and et al. (1993), *Gecekondularda Ailelerarası Geleneksel Dayanışmanın Çağdaş Organizasyonlara Dönüşümü* (The Development of Inter Family Traditional Solidarity Structure to Modern Organizations), Ankara: T.C. Başbakanlık Kadın ve Sosyal Hizmetler Müsteşarlığı Yayınları, p.332,333.

they are relatives or from close kin. If they come to have acquaintances from other villages or provinces even from other sectarian identities, this is due to the physical proximity of living close together for years as neighbors, which is actually a result of involuntary or contingent location. This will be dealt with in detail in the following sections. As indicated by the research of Gökçe[298] the main means of solidarity in the neighborhood depends on the relations between relatives and *hemşehris*.

Consequently migrants from towns other than the ones mostly represented in the region, usually mentioned incidents of having hired waged laborers in constructing their *gecekondus* rather than having enlisted the help of *hemşehris* in the broader sense of the term. In cases where, because of poverty, they could not employ workers, relatives from the village temporarily came to help and they worked intensively in order to finish construction as quickly as possible according to migrant's accounts. These people mentioned almost no incidence of receiving help from neighbors during the construction activity. In cases where migrants are from villages close to Ankara, they mentioned the daily visits of women in the family to support the efforts, construction activity and workers with supplies and services. These migrants, who can be defined as "outsiders" deriving from common feelings of loneliness and helplessness, made references to the mutual cooperation and help in the early periods of settlement. The "outsiders" who were left as in minority position mainly vis-à-vis migrants from Cemele/Kırşehir helped each other since they shared an identity as neighbors. An old man from a village of Yozgat which was under-represented in the region, explained the difficulties he faced in the initial periods of settlement in the following manner:

> I went to a meeting in my child's school and complained about the laziness of my son to the teacher. He said: "don't say anything bad to the child. Two students are successful in this class and one of them is your child, encourage him". Then I came home... We had no electricity at the time but I thought that I had to get electricity so that my child could work. There was electricity in the neighborhood but we had no money to take it in the house. We had a very good neighbor who is from Çorum. He helped me and gave me electricity from his house with a cable. They told him that the municipality might have punished him for that, but he accepted it. He was really a very good man.[299]

298 Gökçe (1993:297).

299 Extract from filed research note no. 22, 3 May 2003.

However, the nature of help among "outsiders" seemed to be different from the one among close kin or people from the same village. Giving and asking for help was described as less straight forward, help was requested less often and the quality of help was limited to allowing access to tools or domestic resources like using the well or electricity for a certain period of time in return for money. A woman from Kırşehir/Kaman described the ways of obtaining help from her neighbors from other towns in the following manner:

> We had a lady here, a good neighbor. In our first night here, she took my girls and opened her house to them. She was a woman from Koçhisar [a town of Ankara]. We had no running water at the time. I asked their hosepipe. I told her to check her water meter [so that I could repay the cost of the water]. They also gave us a glass of tea. I could never forget this kindness.[300]

During the course of the research the migrants often mentioned the importance of *hemşehrilik*, particularly in relation to the early periods of the settlement. *Hemşehris* at the initial stages of the settlement refer to more intimate associates such as the relatives and villagers who can be considered within the frame of close kin according to the findings of the research.

In a recent study on a *gecekondu* settlement in Ankara, it has been pointed out that *hemşehri*, which once meant "being from the same country of origin", in time took on an extended meaning for *gecekondu* settlers: "being from the same ethnic or sectarian group".[301] This finding holds true also for more recent definition of *hemşehri* in Boğaziçi neighborhood. However, during the initial settlement process and construction of *gecekondu* houses, neither the people from the same province nor people from the same religious sect considered that such identities alone felt obliged them to help each other. This finding of the research is supported by the earliest and one of the most sophisticated *gecekondu* research of Kemal Karpat.[302] He discovered that the scope of and closeness generated by *hemşehrilik* grows in direct proportion to the distance from the original location. "For Turkish villagers and their relatives and *hemşehris* in the city, blood and communal ties provide the only basis for mutual help and solidarity until these are superseded by or fused into other relations and identities of urban and national origin".[303]

---

300 Extract from field research note no. 33, 30 June 2003.

301 See, Ayata (1991).

302 Karpat, Kemal (1976), *The Gecekondu: Rural Migration and Urbanization*, Cambridge University Press.

303 Karpat (1976:85,86).

The lack of water, electricity, and construction materials in the early periods of the settlement placed great pressure on the everyday lives of the migrants. The most important tactic, which was to finish construction as soon as possible, could only be sustained by the unconditional help of such community that was defined in a strict and limited way by the migrants.

4.3.2 Near-Strategical Acts Defining Community Settlement

The second important tactic in *gecekondulu*'s dealing with the system was to settle in suitable places where they could easily prevent their *gecekondus* being demolished and be protected from possible natural disasters. However, migrants made the settlement decision in a hurried way, most of the time under a certain time pressure, making an assessment of the existing material conditions and opportunities available rather than taking future considerations into account.

According to the narratives of the migrants, the available opportunities were mainly shaped by their *hemşehri* predecessors, who, most of the time, provided the newcomers with suitable land to buy or settle on. The decision to construct a house among *hemşehri* community was often voiced by the respondents with an emphasis to the feeling of security during the research. To live within the community made their claims stronger vis-à-vis the system, as believed in by the migrants, therefore this was the most important tactic employed by *gecekondulus* in the initial years of settlement. Second tactical decision is related to the concerns of natural risks. Settlement decisions were apparently effected by calculations which included considering past disasters, as was the case in Boğaziçi. Thirdly, certain tactics were employed by the migrants to appropriate the largest possible plot of land by constructing some part of their *gecekondus* on the land they bought from the villagers and some on state owned lands. Fourthly, as mentioned above, the concerns with being close to affordable transportation services defined the decisions of where to settle.

Many migrants, however, were unable to apply these tactics considering the future. The possibility on part of the migrants to take these decisions was mainly limited by their financial, communal and experiential power. In most cases, one or more of these choices depended not on migrant's "free" choice but on the opportunities and conditions of the time and place. The early migrants and the people from hometowns that were densely represented in the neighborhood could make such tactical decisions, while others seemed to lack the means to consider all of them in the initial settlement period.

In this sense, people from the early comer and densely represented groups Kırşehir / Cemele and Kırıkkale / Halitli, often gave positive references to their communal settlement patterns as an advantage particularly with reference to certain moments of social spatialization such as the insecure environment of the radically politicized space of the late 1970s. When I analyze the socio-spatial settlement patterns in the region, these two communities seem to cluster in a more concentrated manner when compared to the migrants coming from other hometowns. As will be later discussed, this well equipped them with a certain power to resist outside interference as a community. The ontological insecurity felt by migrants as a consequence of the insecure living conditions within the space of strategy and the risks of natural space made communal concentration inevitable.

In this respect, the leading migrants from Kırşehir, having a strong financial position bought large plots of land from Üreğil villagers solely to sell these lands to their *hemşehris* in order to live close to their acquaintances. This tactical act became a near-strategical one with the growing control over space which these migrants obtained. Other migrants, particularly the ones described as "outsiders" above, hardly made such strategic plans, rather they used the existing opportunities in a limited time span. Different *hemşehri* groups replicated the creation of spatial boundaries over time, which had made the dialogue between different *hemşehri* and sectarian groups difficult. Such spontaneous social spatialization altered the balance of power relations relating to the political, economic and social context at the time. The effects were mainly felt in the social space of the neighborhood in the 1970s, that will be addressed in the following chapter.

To sum up, the leading role of the early migrants, relates variously to their early arrival to the region, to their economic power, to the entrepreneur qualities, and to the respect they have in relation to their social network in village or city. The wife of a contractor in her 70's from Kırşehir/Cemele describes her husband's attitude towards his *hemşehris* as follows:

> He bought all the lands from the top of the street up to here. [She refers to the lands on two edges of a street, approximately 100 meters long. There are approximately thirty houses of Kırşheir/Cemele migrants.] Then he sold these lands only to our *hemşehris* so as to make them our neighbors. He bought them from an Üreğil villager, a respected man and sold them to our *hemşehris* from the price he had bought, with no profit. He sold land only to our people from the village.[304]

An important Kırşehir/Cemele community formation in the region seems to be made possible with the efforts of people like the contractor mentioned

304 Extract from field research note no. 41, 19 Aug. 2003.

above or with the attempts of wealthy people in the community. Strict social control seemed to be empowered during the period in order not to permit the entrance of outsiders and their construction *gecekondu* houses in the community area. In almost every decision to settle or migrate to a certain region, the prior settlement of a relative group or members of close community (villagers) have played the defining role. Karpat[305] defines these migrants as *kurucular* (founders) and he states their role in shaping the space of the neighborhood as vital similar to the findings of my research.

Accordingly, the migrants who described their financial sufferings in the early settlement period chose places to settle considering certain parameters, like the cost of the land, conditions of purchase and *hemşehri* links and help. In that sense, particularly the migrants from certain villages of Ankara, Yozgat, Çorum and Çankırı who emphasized their poverty at the time, preferred to settle in places where land was rather cheap. The land values were very much affected by the considerations of natural factors in the early settlement period. As mentioned above, the natural space and its appropriation in certain ways had constituted the main concern for migrants in those years.

The main component of nature in the late 1950s and early 1960s which made the most lasting impression in the memories of the migrants is the river and related stories. The migrants describe the river that ran through a valley cutting the neighborhood into two parts as one of the most important natural reference point in their everyday lives. The migrant narratives of the early physical conditions of the neighborhood frequently begin with a highly emotional description of river and the several incidents related to its flooding. The river flooding, a primary concern at the time, greatly influenced the course of spatial development in the neighborhood. River flooding prevented migrants settling near the river and this risk perception led to a decline in the value of lands either side of it. As understood from the narratives of the respondents, the decision about where to settle was taken mainly with reference to their positioning vis-à-vis the river. The lands adjoining and immediately above it were the old vineyard areas, which were mainly private property. The lands that were closer to the top of the two hills were mostly state-owned, and were the most topographically unsuitable and difficult lands in the region to settle. In the late 1950s and early 1960s, the least valuable lands in the region were the ones near the river due to the often-encountered risk of river flooding. Therefore less prosperous migrants in the 1960s relate their preference for either the lands adjacent to the river or the state owned lands on top of the hills. This was

305 Karpat (1976:80).

due to the free availability of the naturally inferior quality, and therefore low value, state owned lands.

The vineyards in between were the most valuable land and were regarded as both naturally "safe" and easy to settle in the early period, and were therefore appropriated by wealthier migrants from Kırşehir and Kırıkkale. The early settlement pattern in the neighborhood was influenced by all these factors, mainly related to the limitations and opportunities of natural space, which emerged from the stories of the respondents. These near-strategical acts of relatively powerful groups in avoiding settlement near the river, however became a source of serious regret, because, as they described in their own words over the coming decades, the value of this ideal settlement land began to lag behind the value of land closer to the river following the construction of a market and the subsequent remarkable increases in land values there. As mentioned in the previous chapter, the land of the region exhibits a mixed structure in landownership patterns. The planning practices and the municipal interventions of municipality from top to down also played a role in the rapidly changing and unexpected value increases, apparently contradicting the early and near-strategic estimates of well to do migrants, which will be dealt with in the following chapter.

Thirdly, the size and nature of plots that had been selected might incorporate some tactical decisions to utilize the loopholes in the landownership patterns. As mentioned above, first wave migrants mainly preferred to buy privately owned lands of Üreğil villagers rather than the state owned lands on the hilly parts. State owned lands on the upper parts of the hill topographically exhibited undeserved qualities for settling on. However, there were also state owned lands left in-between the farms and vineyards of the Üreğil villagers. Migrants who purchased lands from the villagers considered the location of these state owned lands vis-à-vis the land they bought, related to their desire of extending their rights to the land by appropriating the state owned lands near the plots they had bought. They mentioned their desire to build *gecekondus* on such lands that show a pattern of mixed ownership in order to legalize the state owned lands appropriated and add them to their property under amnesty laws following subsequent elections. Thus, some migrants constructed their houses not on the land to which they had legal title deeds but on these nearby state-owned lands. This officially-owned land constitutes the garden of their *gecekondus*.[306] Migrants knew that the next amnesty law to be established would legalize this unlawful appropriation for the sake of the *gecekondu* house constructed

306 This act can be considered as a tactic of "colonizing space" that seems to transcend the limits of De Certeauan definition of tactic.

on it. In this way they would extend the size of their land by paying a token fee to the state:

> We encouraged our father to buy land here because everybody from the village had already bought land in Boğaziçi. He did so. Then we came and bought briquette. We constructed the house mainly on the state owned parts of the land, a little bit behind by paying attention. We left the land with title deed (privately owned land) as garden of our *gecekondu*.[307]

Migrants were aware of the fact that if they could preserve their *gecekondus* until the approaching election, the amnesty law would legalize them. Their near-strategical acts involve the perceptive tactic of taking into account that unlawful acts tend to be overlooked by the strategy of urban policy.

4.4 The Story of the River: The Beginning of the Journey from Natural Space to Market Place

The river that once ran all the way through the valley at the center of the neighborhood occupies the primary place in the memories and narratives of the migrants. Particularly, migrant's early memories of the neighborhood include certain references to the river and incidents related to it. The nature of these memories changes, depending on the gender, age, or economic well-being of the migrant. The river plays an important role particularly in the reminiscence of older women, migrants who were children at the time being and poor families who had settled near river.

In the late 1950s and early 1960s, river had constituted a place to socialize for women while washing clothes there. They describe singing folk songs together, drinking tea or having picnics near the river. As understood from their narratives, spatial relations in the village were replicated in the urban setting in the first settlement years, mainly due to the resemblance between *gecekondu* space and village. The primordial characteristics of the two spaces, the resources they provide and the qualities they maintain, which define the relationship between women and space and the definition of work, were very much the same. For the migrants, the river provided many resources in the early settlement periods, such as providing water and sand for the construction of *gecekondu* houses. The migrants not only used the sand for house construction, some poor families also sold it to the factories nearby as

307 Extract from the field research no. 29, 20 June 2003.

a source of income. A shop owner form Çorum described the poverty they had experienced during the initial years of settlement and their use of river as follows:

> I went to junior high school in Mamak. Though the bus ticket was 25 *kuruş* at the time, I often could not afford it with my pocket money. Therefore I often walked from the school to Boğaziçi [the center of Mamak is approximately five kilometers a way from the neighborhood]. I shared one pair of shoes with my brother in those years. On one occasion, my father could not afford to buy me uniform and was very upset. I pulled sand from river that had been washed up near our house after flooding and sold it in barrels. Ooo, I made a lot of money out of that [he laughs] and bought the school uniform myself that year.[308]

The river, as the most important provider of natural resources for construction such as sand and water, had been a spatial focal point in the lives of the migrants through the 1960s. Their survival tactics included the essential and determined exploitation of natural resources, leading to the construction of small workshops and factories near the river. These small workshops including ironworks, sand factories, hardware stores and lumber merchants were described as the prototype components of the market place. These workshops were vital for the migrants due to the continuous construction activity in the region, either in the form of constructing new *gecekondus* or making improvements and additions to houses already built. The necessities of daily life and their unique relationship with natural space determined the course of development and social space in the region during these early years of settlement.

As mentioned before, in the memories of the migrants, the river had intimidating effects due to the risk of flooding. Due to this risk, the plots on either side of the river were least valuable when compared to privately owned lands in other parts of the neighborhood in the early years of the settlement. The first known flooding took place on September 11th of 1957 in Hatip River and its branches all over Mamak due to heavy hail. At its height, the torrent reached three meters in Kayaş and its effects were felt everywhere between Mamak and Kayaş. This was a real disaster, which resulted with the death of 133 migrants.[309] *Zafer* and *Ulus* daily newspapers -supporters of Democrat Party and Republican People's Party respectively- initiated a

---

308 Extract from field research note no. 7, 16 Oct. 2002.

309 Şenol-Cantek (2006:59).

public debate about this incidence. After the flooding had taken place, the governor of Ankara claimed that the improvement and demolition policies of *DP* had dramatically reduced the potential loss caused by flooding. The closure of the upper part of the river due to the improvement plans of *DP* was said to have lessened the flood damage.[310] The daily *Ulus*, on the other hand, severely criticized these claims of the governor in a public debate highly critical of "despotically modernist" and formalist policies of *DP*. *Ulus* represented the declarations of governor as an attempt to legitimize and facilitate the predicted demolition orders by *DP*.[311] It is important to note here that *RPP*, a party that was once the ardent supporter of modernist urban policies criticized the demolition decisions of *DP*. *DP*, on the other hand, unlike its early populist approach to *gecekondulus*, seemed to internalize rather a modernist and formalist approach in these years with regard to *gecekondu* settlements. This case clearly reflects the hesitant approach of all political parties with regard to *gecekondu* settlements at a time when the political and economic context was changing.

The flooding did not cause much harm in Boğaziçi, since the river here was only a small branch of Hatip River. However, it made an impression on the memories of the migrants. In most cases, poor families lacking close and strong *hemşehri* ties and networks had settled lands nearby river, while well to do migrants from Kırşehir/Cemele and Kırıkkale/Halitli preferred safer places. An old Boğaziçi *muhtar*, originally from Kırıkkale/Halitli, defined their settlement history as a *hemşehri* community in the following manner:

> An old man from our village had come here first. He had a house nearby river in the village and he had suffered from flooding so much there. Therefore, he preferred upper parts in order to settle. That is the reason why we are all here now.[312]

An old man from Kırşehir expressed their settlement decisions in a very similar manner as follows:

> When we first come here, we thought to get a land near river. There was a man who wanted to sell a plot. He wanted 825 *kuruş* [he refers to the low values of land nearby river] for each square meter. At that time a flooding occurred and your aunt [referring to his wife] was so scared and we didn't buy the land. Now it is

310 Ibid., p.58.

311 Ibid., p.59.

312 Extract from field research note no. 29, 20 June 2003.

> so difficult for us to climb that hill. If we wanted to make some alternations to the house, we couldn't carry sand and construction materials up here.[313]

The settlement decisions of migrants from Kırşehir have constituted their main regrets in life in the following years with the unexpected land value increases near the river. The lower-income families, on the other hand, could afford to buy these places not only because the land values there were lower but also influential actors had played the creditor and mediator role in the exchange of these lands between Üreğil villagers and migrants.

One of the most important actors was "family X" from a village of Çankırı. Since the family is from Çankırı, they can be considered in the category of "outsiders" vis-à-vis the migrants of Kırşehir. The power of the family, in that sense, did not emerge from the support they got from their *hemşehri* community in the region. However, they had a peculiar position and were respected by migrants even from Kırşehir for a number of reasons emerging from their settlement history in the region.

They were one of the poorest families in the neighborhood in the early 1960s. When they first migrated to Ankara as three brothers, they began to work in a bakery in Hamamönü. The owner of the bakery offered them a room to stay. Over time, they began to sell bread and small subsistence goods to *gecekondu* neighborhoods in the periphery of the city on a horse coach in the 1950s, one of which was Boğaziçi. Every migrant in Boğaziçi, mostly *Kırşehirlis* knew them at the time due to their supply of acutely needed subsistence goods. Subsequently, they opened a small *bakkal* in the neighborhood near the river due to the low cost and the central position of the lands there, as recalled by a member of the family. On the shop, they built one extra room which they lived in. One member of the family mentioned the unconditional help shown by the neighbors to them during the construction of their *gecekondu bakkal* in the following manner:

> We bought materials from the warehouse and we dried bricks. Everybody helped. We also worked at nights during the construction. Once a watchman came and he got angry with us, because we did construction at night. We had a neighbor from Kırşehir. He told the watchmen not to pick on us. Women from the houses around came and they asked us what our needs were: "Mud bricks and water, or pickaxe and shovel?" At that

313 Extract from field research note no. 36, 20 July 2003.

time everything was problem; water was a problem, there were privately owned wells everywhere.[314]

The family had been given great assistance in their attempts to build a *gecekondu bakkal,* even though they were outsiders vis-à-vis extended *hemşehri* groups deviating from the general pattern. As mentioned above, this seemed remarkable, considering the fact that people were generally unwilling to offer unconditional help to neighbors unless from the same village. This support can be considered a consequence of the provision of essential services offered to migrants in the early period of the settlement by this particular family, a point repeatedly raised by respondents during the interviews. In addition, there was only one *bakkal* during those years in the neighborhood. As stated by one member of the family, the migrants were largely favourable to the construction of a second *bakkal* due to the desperate need of basic subsistence goods in the neighborhood in the late 1950s. In spite of the initial poverty this family experienced during their early years in city, as time passed, due to their accurate perception of the needs of the neighborhood they had opened other shops and even two cinemas in the neighborhood by the 1970s.

After having earned certain capital from their *bakkal,* they bought farms and vineyards of Üreğil villagers in big plots, or appropriated state owned lands and built houses, either adjacent or in close proximity to the river for the newcomer migrants who were in need. They sold *gecekondu* houses to the poor in installments. During the interview they described themselves as being in the real estate business. However, their influence was much more than an ordinary estate agent because they managed and controlled the social space of the region by selling land and house to certain chosen people. According to the narratives of the migrants who bought land from the family, it became obvious during the course of *gecekondu* legalization in the late 1980s that the family did not, in fact, own all the lands they had sold to the migrants. When we consider the fact that the lands of the region reflected a mixed nature of private and state owned lands, this family seemed to use the state-owned extension of the land they bought from the villagers in order to build and sell *gecekondus* and shops. The shop owners from the villages of Çankırı, Çorum, Yozgat, Kayseri or Ankara, particularly who are now considerably richer when compared to other shop owners, generally mentioned their purchase of land or *gecekondu shop* on credit from "family X". This made the social scene of the market more heterogeneous. This family seemed to contribute to the value increases of lands near river by

314 Extract from field research note no. 16, 8 Apr. 2003.

means of helping poor migrants open a number of shops. The emergence and growth of commercial area in turn contributed to their wealth and political connections, strengthening their position over time. The construction of a big market place in the region was also due to this family constituting an example to the newcomer migrants and their help to the lower-income entrepreneurs. A wealthy shop owner from Ankara/Kalecik who owned one of the five apartment houses in the market place describes the help of the family during their early settlement period. As confirmed by this shop owner, this family not only provided credit opportunities in the purchase of their shop but also provided guidance on the trustworthiness of various customers. In fact, the family seemed to serve as "landlords" in *gecekondu*, as defined by Şenyapılı.[315] These landlords usually appropriated state owned lands or purchased others and sold them to the newcomers. On some occasions, they built houses and sold these completed houses on credit. In confirmity with the general pattern, "family X" also opened warehouse shops where they sold construction materials, in addition to their role as a mediator for the new comer migrants by providing entrance to the region. They offered financial incentives to those taking several of their services.

The river and its transformation from natural space to a commercial center seem to deserve attention in the understanding of the the spontaneous spatial development in the region and unexpected consequences of near-strategic acts on part of powerful actors like "family X". Another cause of the rapid development of market place was the transformation of the riverbed, and its branches flowing from the top of the hills, into the main routes of migrants travelling to and from the railway station. The traditional route following river became, over time, a main artery.

It was as a result of these factors, rather than the predictions and near-strategic acts of *Kırşehirlis* in the initial period of the settlement that values of lands near the river increased considerably. This constitutes one of the most important missed opportunities in the lives of *Kırşehirli* migrants. Though, migrants from Kırşehir constituted the strongest group in economic terms in the early settlement period, they merely watched while lands on either side of the river were seized by "outsider" migrants. Therefore they were unable to have much influence in the shaping of the public space of the neighborhood. Most of the shop owners of Kırşehir origin bought or rented shops in the market area in the 1980s at price reflecting the increased value of real estate. Their authority in controlling the development of public space was negligible when compared to their numerical presence in the neighborhood. On the other hand, less well of migrants, who suffered

315 Şenyapılı (2004:193).

poverty in the early periods, from villages near Ankara, such as Çorum, Çankırı and Yozgat, gained economic power because they built houses and started businesses near the river. The near-strategical acts of "family X" constituted the main reason behind these people's gaining power in time.

As will be explained in the following chapters, the near-strategical acts of "family X" gave them a certain control over the shaping of social space of the region. The family played the role of the most influential agent in the development of the commercial area by controlling the purchase of land and construction and by guiding the newly arriving shop owners in terms of social relations as stated by many shop owners.

The story of an *Alevi* shop owner explicates the family's attempts to control the social spatialization in the region. He mentioned how his father had struggled to take possession of a house and a shop near the river and how other shop owners ostracized them as a result of their struggle with "family X". At present there are only four or five *Alevi* shops in the market place, composed of nearly 150 shops. The general exclusion from the most valuable and central lands of the region of more recently arrived *Alevi* families from Çorum, Sivas and Tunceli due to their sectarian identity and the timing of their migration would have spatial consequences in the subsequent decades, and will be dealt with in the following chapters.

4.5 Conclusion

Chapter four is an attempt to read the spatial development of Boğaziçi neighborhood in the early settlement period. I have tried to assess the relationship between the strategical realm, the realm of "modern", "urban space" as represented and dominated by the planners and the national and local urban policies, and the tactical realm, the realm of migrants' acts in appropriating the space of nature. The narratives of migrants relating to the initial settlement years in the region have made it possible for me to read the interconnected and dialectical histories of "strategy" and "tactic". This dialectical history referring to the case of Boğaziçi necessitates a critical review of the theories of Lefebvre and De Certeau and a reconsideration of the power of strategy and tactic vis-à-vis each other.

A horizontal model of the power relation between migrants has added to the vertical model, considering the limitations of the latter. On the horizontal level the tactical acts of migrants vis-à-vis each other and vis-à-vis the space of strategy on the vertical level seem to signify a dynamic model. The dynamism in the model and the spatial contingencies that led to unintended consequences seem to constitute the main reason behind the spontaneous development of the region.

The spatial contingencies emerging out of the weakness of the planning and urbanization strategies of the Republican Regime in Ankara were considered with reference to the survival tactics of migrants in Boğaziçi and the spontaneous development of the neighborhood. The impediments of strategy in making future projections about the spontaneous spatial developments, and the tactical acts of migrants and state officials within the loopholes of the system, were exemplified by the incidents from the lives of Boğaziçi dwellers.

The tactical decisions of early comer migrants seem to transcend the definition of tactic by De Certeau. These "tactics" on the part of powerful migrants can be defined more properly as near-strategic acts of colonizing neighborhood space. In this early period, the natural determinants constituted the context of the power relations on the horizontal level. However, similar to the vertical model of power relations, the near-strategic acts would lead to unintended consequences for these powerful groups, like the change of real estate values due to certain spatial factors in the early settlement period. The shaping of the social space of the neighborhood by these near-strategical acts on the horizontal, and strategical acts on the vertical level, seems to have contributed to the dynamic nature of power relations. In the 1960s the early-comer migrants from Kırşehir / Cemele, Kırıkkale / Halitli and "family X" seem to have operated on the near-strategical level in shaping the social space of the region. However, the unintended consequences of these acts would necessitate a reconsideration of the nature of social spatialization in the subsequent decades.

CHAPTER 5 - The 1970s: The Impacts of Radical Politics on Social Spatialization in the Locality

5.1 Introduction

This chapter seeks to narrate the story of social spatialization in the neighborhood in the 1970s, where the public space of the neighborhood was radically politicized. It also examines the consequences of this political transformation. As mentioned in the previous chapter, the tactics employed by the migrants on the horizontal and vertical level vis-à-vis each other, and the interventions of strategical agents gave social spatialization in Boğaziçi a spontaneous quality.

The near-strategical acts of the founders in the 1960s led to the exclusion of certain groups from the most critical aspects of neighborhood space near the river basin gained considerable value over time. This exclusion affected the course of spatialization in the late 1960s and 1970s, where the public space was radically politicized due to the macro economic and political developments in the country. With the radical politicization of space, the groups who had some power to define the course of social spatialization in the previous decade came under pressure from the previously excluded groups. This was mostly due to the fact that the spontaneity had reached its peak in terms of its impacts on social spatialization in the locality during the 1970s. The limited capacity and the political bias of the state in intervening in the sphere of the tactical realm added to the degree of spontaneity. In the 1970s the power relations and the radically political struggles between militant groups seemed to define the context of social spatialization in the locality almost in isolation. This attributed certain autonomy and peculiarity to *gecekondu* spaces, differing from the middle and upper-middle class districts of the city during the period. The nature of social spatialization in the neighborhood in the 1970s will be evaluated with such theoretical concerns in this chapter.

The respondents describe the period with a strong emotional depiction of the political struggles which took place in the public space of the neighborhood. As different from the narratives of the previous decade, there emerges an implicit reference to the identities of *Alevism* and *Sünnism* in order to denote "us" and "them". This was made explicit with reference to the political identities of being leftist and rightist. The *Sünni* or *Alevi* respondents in their narratives regarding the 1970s do not approve of the struggles between young leftist and rightist militants in the public space of the neighborhood. Yet again in their narratives, this period signifies the moment where they initially located themselves and their *hemşehri* community within these extended political identities of the left and right, associated with *Alevism* and *Sünnism* respectively. In this sense, in the memories of the inhabitants this period in all cases almost is narrated with reference to the politically overloaded identities of *Alevism* and *Sünnism*. The politicization of sectarian identities and, as a consequence, the *hemşehri* identifications were closely related to the macro political developments of Turkey at the time.

5.2. An Overview of the Political Context in the late 1960s and 1970s

The 1960 military coup against the *DP* rule paved the way for two decades of economic and political crisis in the history of the country. The 1961 constitution, after the military coup, included the bill of civil liberties prepared by a commission of university professors. In political terms the opportunities the constitution provided were only fully exploited from the mid-1960s onwards.[316] Political debating societies (*Fikir Klüpleri*) sprang up at all the major universities.[317] However, the liberty of the period, in terms of the opportunity for diverse groups to express themselves, when combined with the economic crisis and the increasing rate of rural-to-urban migration with *gecekondu* neighborhoods in cities becoming ethnically more diverse spaces, led to the surfacing of competing claims. The political landscape of the later 1960s showed extreme diversity, with parties and illegal groups proliferating on the extreme left and right of the political continuum while the two major parties positioned themselves on the right and left of center. These were the *JP* (Justice Party) and the *RPP* respectively.

The *JP* under the leadership of Demirel, won the 1965 parliamentary election with a 52.9% majority. As stated by Zürcher[318] Demirel, to keep the party's integrity with a very wide range of social groups (coalition of industrialists, small traders and artisans, peasants and large landowners, religious reactionaries and Western oriented liberals) suggested two strategies. Zürcher defines these policies as tactics. However, within the framework of this study, these policies are considered, conceptually, macro *strategies* of the Demirel government.

First, he emphasized the Islamic character of the party and the way it stood for traditional values, and second, he kept up a constant campaign of anti-communist propaganda and of harassment of leftist movements.[319] These two tactics seem to reflect the general approach and affiliations

316 See Zürcher (1998:257,266).

317 Ibid., p.268.

318 Ibid., p.263,264.

319 Ibid., p.263,264.

of the state in dealing with the politically radical atmosphere of the late 1960s and 1970s. As mentioned in the second chapter, the combination of a persistent balance-of-payments deficit, and an industry that was dependent on foreign capital, made the Turkish economy extremely vulnerable. More than the social unrest, or even the violence in the streets, it was the growing economic crisis, which defined the governments of the later 1970s.[320]

During these years there was a struggle between rightist and leftist militants in order to control certain spaces in the city. However, as considered by many, the struggle between right and left was an unequal one. The police and the security forces had become advocates of Türkeş's *MHP* (Nationalist Action Party) during the "Nationalist Front" governments of the years between 1974 and 1977. "Even under Ecevit's government (the party of center left), state forces remained heavily infiltrated by fascists who shielded and protected the *Grey Wolves* (the youth organizations of ultra nationalists) in 1978 and 1979."[321] The growing repression of leftist groups by ultra nationalist rightist groups, and state security forces, which mostly associated themselves with nationalist ideology, paved the way for political polarization and radicalism in the country. Despite the fact that in 1950 the *Alevis* supported the *DP* as a reaction to the authoritarian and centralist policies of the *RPP*, by 1957 they had given their support to the *RPP* again. In the late 1960s, *Alevi* youths were, to a large extent, associated with socialism and extreme leftist politics. *Gecekondulus* began to vote heavily for leftist parties and *gecekondu* spaces gained an oppositional position to the system.[322] Meanwhile, the ultra-nationalists mobilized the *Sünni* population around the *MHP*, with policies that equated anti-communism with anti-*Alevism*.[323] As will be mentioned in the following pages, such identification of communism with *Alevism* turned the identities of *Alevism* and *Sünnism* into conflict-ridden identities, even for the majority of *gecekondulus* who did not hold any active and obvious political position.

In congruence with the general context of the country, the public space of the Boğaziçi neighborhood is narrated as the arena of competition and struggles between leftist and rightist groups in the individual stories of the migrants. The developments brought about the politicization of public

---

320 Ibid., p.280.

321 Ibid., p.276

322 See Aslan, Şükrü (2004), *1 Mayıs Mahallesi: 1980 Öncesi Toplumsal Mücadeleler ve Kent* (May 1 Neighborhood: Pre-1980 Social Movements and City), İstanbul: İletişim, p.81.

323 See Göner, Özlem (2005), "The Transformation of the *Alevi* Collective Identity," *Cultural Dynamics*, 17(2), p.110.

space were closely related to the general political context of the country, as mentioned above. However, the way the political struggles between groups took place in *gecekondu* neighborhoods involved a certain form of spontaneity. The struggles between extreme militant groups seemed to take place in a more direct and harsher manner in these neighborhoods. In his evaluation of the 1970s, Zürcher pointed out that traditional culture, in which honor and shame and an extreme contrast between one's family or clan and outsiders, made the political extremism in Turkey so exceptionally violent.[324] Traditional conflicts seemed to be taken as political connotations in *gecekondu* settlements, and in some cities where the living spaces of *Alevi* and *Sünni* populations were close. The most serious and bloody clashes between *Alevi* and *Sünni* groups took place in anti-*Alevi* pogroms in the provinces of Malatya, Kahramanmaraş and Çorum.[325] The Maraş Massacre of 1978, in which 109 people were killed, was directed against *Alevi* neighborhoods in the city of Maraş.[326] The spontaneity of these settlements, differentiating them from other neighborhoods of the city, was related to this condition of heterogeneity in terms of sectarian identities. In most cases, this had contributed to the negative image of *gecekondu* neighborhoods in the public debate, as potential shelters for illegal and conflict-ridden identities and groups. As mentioned in the second chapter, this perception in public discourse that emerged in the late 1970s has continued its existence by taking different forms until the present.

As was mentioned in the previous chapter and will be exemplified in the following chapter, the definition of sectarian or hometown identities in the narratives of the respondents showed variations, depending on the social context and time. The nature of the struggles between leftist and rightist groups also shows variations from one *gecekondu* neighborhood to another. In this sense, the narratives about the neighborhood of the 1970s will hopefully give clues about the reasons behind such spontaneous happenings that set these neighborhoods apart, not only from the middle class neighborhoods of the city but also from other *gecekondu* neighborhoods in the city.

Depending on the narratives of the migrants, this was a period where state intervention in neighborhood space seemed to be relatively insignificant in terms of security, supervision and planning when compared to other periods. Particularly, due to the insufficiency and political bias of the state security forces, *gecekondu* neighborhoods (except those dominated by *Alevis*) were left to their own affairs. The struggles between state forces

324 See Zürcher (1998:277).

325 See Bruinessen (1996:8).

326 See Göner (2005:115).

and militants of *Alevi* dominated leftist neighborhoods took a quite violent form. As Zürcher mentioned, the authorities were not able to restore order in these neighborhoods. In some cases, whole neighborhoods came under the control of one or the other of the competing groups and were declared "liberated areas".[327] However, Boğaziçi was considered as a *Sünni*-dominated neighborhood. In this respect, despite the activities of leftist militants in the locality, it was left, to a large extent, to its own affairs by state forces. This contributed to the spontaneity of the locality.

As mentioned in previous chapters, Boğaziçi was also socially composed of different ethnic groups that were identified mainly through hometown connections. The (socio)-spatial encounters between such groups, and the immunity they had due to the limitations of their strategic power to deal with the tactical acts of *gecekondulu* at the time contributed to the spontaneous spatialization of *gecekondu* settlements. Beginning with the 1970s, the social heterogeneity of the neighborhood further increased with the second wave of migration, mostly from the *Alevi* villages of Çorum and Sivas. In addition, a large group of *Alevi* / Kurd migrants of Tunceli came from another *gecekondu* neighborhood of Ankara (Yenidoğan) to Boğaziçi by way of an in-city, chain form of migration. As Nail Yılmaz[328] mentioned, the *Alevis* who had lived in their villages as closed and homogenous communities and had migrated to the cities later than *Sünnis*, could not find a suitable environment to enact their practices. Because of the obstacles of their peculiar religious practices and because they made their *Alevi* identity explicit, the socio-economic conditions of the *Alevis* were not as well-to-do as the latecomer migrants to the city when compared to *Sünnis*.[329] "Their gradual integration into the wider society, migration to the towns, education, careers in public service brought them in closer contact and sometimes in direct competition, with strict *Sünnis* from whom they had remained socially separated for centuries".[330] As shown in the research of Gökçe et. al (1993), ethnic and sectarian groups, as a result of their latecomer status and minority position, mentioned their conscious preference for close relations only with relatives and the close *hemşehri* community, rather than with neighbors.[331]

Only from the 1961 onwards, *Alevis* founded associations, vaqıfs, cem houses and political parties in order to make their voices heard.[332]

327 See Zürcher (1998:277).

328 Yılmaz, Nail (2005), *Kentin Alevileri* (The *Alevis* of the City), İstanbul: Kitabevi, p.151.

329 Ibid., p.181.

330 Bruinessen (1996:8).

331 Gökçe et. Al (1993:289,292).

332 Yılmaz (2005:181).

Consequently, in the case of Boğaziçi, the increasing heterogeneity of the neighborhood when combined with the socio-spatial consequences of the exclusionary near-strategical acts of the powerful groups in the preceding decade, seem to determine the way the competition between groups took place in the 1970s.

Depending on the narratives of shop owners in relation to the late 1960s and early 1970s, both hometown commonalities and sectarian identities played an important role in the construction and enlargement of the market place. The conflict and uneasiness between the different groups in the enlargement period of the market place in the 1960s, as gathered from the narratives, explain to an extent the ever-expanding politicization of the market place, the most important component of public space in the coming years. In this sense, the powerful groups who had taken near-strategical decisions in the shaping of the market place in the 1960s suffered most from the political struggles of the 1970s.

As mentioned in the previous chapter, there was explicit outsider-outsider solidarity in the market place. As was explained, "family X" from Çankırı was the main actor playing the role of mediator by giving financial and social help for land provision and shop construction, particularly in the narratives of the "outsider" shop owners from Yozgat, Çorum, Çankırı and the villages of Ankara. There were almost no *hemşehri* bonds between this family and the migrants who received their help. As Sema Erder[333] mentioned in her field research findings, there needs to be no direct *hemşehri* relations between the founders and the inhabitants. Thus, the founders in her research setting had no *hemşehris* in the region. The "family X" had a similar status in this research setting by having no *hemşehri* community. As already mentioned in the previous chapter, their acceptance by Boğaziçi inhabitants depended on certain other factors. However, this does not mean that *hemşehri* identities defined through sectarian lines had no effect on the preferences of the family in their provision of certain services to the newcomers. Beginning from the very initial phases of the production of the market place to the early 1970s, the family did not encourage *Alevi* migrants to settle near the river. It is interesting to note that the few *Alevi* shops already existing are located in a place close to the Samsun Highway, the entrance of the neighborhood. In other words, the lands and shops of the family seemed to constitute the boundary between the few *Alevi* shop owners and the other *Sünni* shop owners. Beginning from the lands owned by the family to the center of the market place, there was only one *Alevi* shop. The family seemed to spatially constitute the main barrier to

333 Erder, Sema (1996), *Ümraniye: İstanbul'a Bir Kent Kondu* (Ümraniye: A City was Settled Upon İstanbul), İstanbul: İletişim Yayınları, p.73.

the enlargement of the *Alevi* lands towards *Sünni* dominated places. *Kırşehirli* shop owners and other *Sünni* "outsiders" also did not welcome *Alevi* shop owners. There is no case I encountered where a *Sünni* landowner or a shop owner sold his plot or shop to an *Alevi* person. In a few incidents, some early-comer *Alevi* migrants appropriated the lands near the river and were able to open shops in the market place. During my interviews with *Alevi* shop owners and inhabitants, nevertheless, there was no explicit reference to the reactionary attitudes of the *Sünnis* during the settlement process. Yet they often mentioned the disadvantageous position of *Alevis* vis-à-vis *Sünnis* with respect to settlement conditions.

The entrance of the market place is not of great valuable when compared to other lands in the market because flooding had caused serious damage to that part. The *Alevi* inhabitants, due to their position as latecomers, resided in the most topographically disadvantageous parts of the neighborhood. *Alevi* shop owners, both as a consequence of their latecomer position and because of the resistance of *Sünni* shop owners, could not settle places close to the center of the market place and the exceptionally rare representation of *Alevi* shop owners in the market place appears to support this. It is also possible to identify clues about the sufferings of *Alevi* shop owners embedded in the migrants' narratives portraying the scene of the market place in the early 1970s. Some shops owned by Kurds or *Alevis* were counted as present in the market in the late 1960s and early 1970s, but they are reported as having then closed. When I questioned the reasons for their closure, as perceived by the respondents, a *Sünni* woman[334], as an example, reported that a Kurd / *Alevi bakkal* left because they could not stand the attitudes of other shop owners. Some others mentioned the difficulties of the *Alevis'* presence in the market where the majority of shop owners are *Sünni*.

Unexpectedly, the near-strategic attempts in the late 1970s of the founder migrants to control the social space of the market and the general discomfort of *Sünni* shop owners with respect to *Alevis* could not prevent the place to turn into the fortress of leftists. Indeed, these exclusionary acts seem to have had a close relation with the fact that these founding shop owners mention their sufferings most in relation to the years of political turmoil when compared to other shop owners. This may also explain the willingness of leftist militants to control the region during the period of political unrest in addition to other socio-spatial reasons. One of these reasons is that the spatial push of the *Alevis* towards the area close to Samsun Highway instigated the close encounters between the *Alevis* and the Üreğil neighborhood, the space of extreme rightist and ultra-nationalist militants. Such positioning of the

334 Extract from field research note no. 39, 15 Aug. 2003.

Alevis caused the train station, the nearby Samsun Highway and the market place to become the main grounds for the political fights and violence, which will be dealt in detail in the following pages.

5.3. The Story of the River: The Expansion of the Market Place and Its Positioning as the Focal Point of Political Struggles in the 1970s

The "bridge" on the Samsun Highway, constructed for the purpose of crossing the river flowing from Boğaziçi to Üreğil, is signified as a symbol of the boundary between the Boğaziçi and Üreğil neighborhoods in the definitions of the migrants. The narration of the neighborhood in the 1970s in almost all cases includes a reference to the bridge separating the political spaces of the extreme right and left in socio-spatial terms.

On the other side of the bridge on the Samsun Highway, in the settlement of Üreğil, close to the train station, most of the residents were *Sünni* in origin. The place reflects a more homogenous social composition, when compared to Boğaziçi. The extreme nationalist militant groups controlled the public space there in the late 1970s. In contrast, the leftist groups, due to dense presence of *Alevi* population in the settlements close by Boğaziçi neighborhood, dominated the market place in Boğaziçi. Though, the market space and the places close to it were mainly occupied by migrants from Kırşehir, the *Alevis* were settled densely on the hills, on both sides of the road.

As discussed in the previous chapter, natural factors determined most of the decisions of the early migrants. The physical form of the riverbed constituted the main reference point for the migrants in their settlement process. The walking paths, following the river and its branches that run from the hills down to the valley, were transformed into the main roads and streets in the following years. The presence of the river in the middle of the neighborhood and its proximity to Samsun Highway made the paths on the two sides of the river the main route used by many people during the day, connecting the distant neighborhoods with the railway station. The only way to the train station passed through the market place, and people living in every neighborhood within the approximate one-kilometer area where the research was conducted have reported using this route during the 1960's and 70's. Particularly in the 1970s when the population of the region and their rate of employment in jobs in the state sector in city center had increased, the market place became the predominant part of the public space. In the old days, the residents of the neighborhoods in the upper parts of the region, mainly inhabited by *Alevi* residents, had to pass the market twice a day, according to the reports of the shop owners and inhabitants of the neighborhood.

Minibuses primarily served the Samsun Highway in the 1970s. However, as often mentioned by the respondents, very few *dolmuşes* entered the neighborhood and upper parts of the region during that period. Depending on the narratives of the respondents, people preferred not to take these *dolmuşes* since they did not feel comfortable for two reasons. Mainly women mentioned their hesitancy and anxiety about using them because of the position of drivers as total *"strangers"* to them. These distant social encounters between the migrants and drivers led to the emergence of concerns related to the security and cost of transportation on the part of the migrants. Migrants mentioned their early uneasiness in relation to *dolmuşes* mainly with reference to the undefined and informal relationships between passengers and operators of the *dolmuş* transportation as compared to the train. As indicated by a shopowner[335], *dolmuş* owners, at times hired drivers from the locality as a tactic to sustain intimacy and acquaintance.

Secondly, people had concerns about the cost of transportation with *dolmuşes*. The train, at the time, was mentioned by all to be to be cheapest means of transportation. Due to the train's status as the predominant means of transportation, *Alevi* people residing in the upper parts of the neighborhood also shopped from the market. This is not the case now, with the expanding possibilities of transportation and shopping in different locations nearby. In this sense, many shop owners described the period of the early 1970s with longing, remembering it as the liveliest period of the market when their earnings were high. An old shop owner with a warehouse narrated the period as follows:

> In the past [referring to the late 1960 and early 70s] this place was livelier than any other place. It was very good... Everybody got off the train close to the bridge. When minibuses had increased in number, they began to go to the upper places. Now a person can go from the front of his house directly to Ulus by minibus. In a case where we are willing accept to sell things on credit or if there is an emergency need, they come and shop from here.[336]

A similar yearning for the good old days was observed during the informal conversations with many shop owners; particularly those who experienced serious economic difficulty at the time referred to these years. Ironically, when the shop owners were asked about their most depressing years in the region, they mentioned 1978 and 1979, when the political struggles on the streets reached their peak. The intervening five or six years had changed the political and social atmosphere of the neighborhood dramatically.

As will be mentioned in detail in the following chapter, the use of the market place by the residents of the closer areas lost its intensity due to a

---

335 Extract from the field research note no. 34, 3 July 2003.

336 Extract from the field research note no. 30, 25 June 2003.

number of reasons. Firstly, with improved transportation facilities, the market place lost its status as the only available route to the railway. Secondly, the Nato Highway had emerged as an alternative route to the city center, serving the needs of Boğaziçi and residents of other neighborhoods living on the upper parts of the hills. Thirdly, after the political turmoil of the 1970s, the identities of *Alevism* and *Sünnism* had been made explicit in defining "us" and "them" in a sharper and exclusionary manner. This further discouraged *Alevi* residents from using the *Sünni* dominated market place.

5.4. The Struggles between Leftists and Rightists Dominating the Social Spatialization

It is important to recall that those shop owners who mentioned their sufferings most in the late 1970s were the ones socially and spatially close to the family of X. They point out the main reason behind their sufferings as their identification with the rightist ideology, either as a result of their hometowns' identification with ultra-nationalist politics or because of the political stand of the people who visited their shops. These shop owners related their experiences about these years as the most depressing ones in their individual histories. Most of them, including the member of "family X", explained their identification with rightist politics as a false impression on part of the leftists. However, during the interviews, they often gave clues about their political standing and their negative approach to leftists in those years. A number of respondents mentioned incidents where the shops of the family were bombed several times in the late 1970s. A member of the family also brought up the incidents of bombing during his interview, and he related the issue to the identification of their shop space as the meeting place of rightists, similar to the perception of their neighbors in relation to the incident.

> The most depressed time was the anarchy period. We went everywhere together as brothers. They [the leftists] bombed here eight times; two of them were one after another. The reason for that... For example, you come here and we chat... Because of people, coming in and going out of the shop... "Why did you talk to these people and let them stay at the shop?" That was their concern...[337]

There is almost no reference to the state security forces and their intervening in the incidents and fights when they happened to take place, according to the interviews and informal chats with the respondents. In this sense, inhabitants narrated certain tactics they employed to sustain their security vis-à-vis the pressures of militant groups in the late 1970s. A gendarme station opened

337 Extract from field reserch note no. 16, 8 April 2003.

in the garden of the primary school in 1975 at the center of the market. However, as understood from the stories, they seemed to lack the ability to sustain security. The shop owners who had their shops close to the center or close to the gendarme station reported that no serious incident took place after the gendarme station was established. However, other shop owners and inhabitants of the region mentioned their sufferings and their ways of solving the problems rather than recieving help from security forces. Related to the bombing, the member of "family X" mentioned the incapability of the commander in the neighborhood in the following manner:

> They bombed here. We called the police; there was a gendarme station here at the time. No one came from this station. Soldiers arrived from Dutluk neighborhood. The Commander asked me where I was from. "I am from Çankırı", I said. He said: "You are a Muslim person I understand". He was a Muslim person too, I understood from his manner [He refers to the commander's rightist political inclinations when he refers to his Muslim identity]. The commander said: "You seem to be thinking going back to Çankırı. [The commander seems to refer to the attitudes of the family's feelings of dread in relation to the pressures of leftist militants.] But do not do that... Resist." He shouted at the commander in our neighborhood: "How dare you could stay there without hearing the bomb?" Next day, military intervention took place. Then we took a deep breath.[338]

As mentioned in the previous pages, the state security forces could not be identified as being politically neutral during the period. Particularly among the police forces, there was an explicit affinity with ultra-nationalist rightist politics.[339] In this sense, the neighborhoods mostly constituted of *Alevi*

---

338 Extract from field research note no. 16, 8 April 2003.

339 Aslan (2004:132) by referring to his monographic study of a neighborhood that was built as part of "*gecekondu* movement" gave examples to the affinity between some ultra-nationalist civilians and police forces. 1 Mayıs neighborhood was constructed as part of leftist class politics by mostly *Alevi* migrants under the belief that poor classes had the right to appropriate public lands for public purpose. See Aslan (2004:79). These neighborhoods were defined as "rescued regions" and subject to continuous demolitions and violent incident. See Ibid., p.85. Despite continuous demolitions, they built their houses again and again. In that fashion, 1 May neighborhood also witnessed a huge demolition by state security forces in which a number of migrants were wounded or killed. During the events, many migrants witnessed the presence of civilians who had put on the rosette of Grey Wolves and act on behalf of and with the police against neighborhood inhabitants.

residents were subject to a high degree of repression by state security forces. As can be gathered from the interviews in Mamak Municipality, Tuzluçayır neighborhood (a region very close to Boğaziçi due to its identification with leftist politics and its inhabitation of *Alevi* migrants) inhabited continuous war-like struggles between the security forces and the leftist militants. In this sense, state interference appeared to be high in such neighborhoods. As previously stated, Boğaziçi neighborhood exhibited a peculiarity in that sense. The neighborhood is heterogeneous in terms of hometown, sectarian and ethnic identities. It is also known as a religiously conservative neighborhood, due to the dense presence of *Sünnis* from Kırşehir and Kırıkkale. In this respect, though the most critical points of public space were under the control of the leftists during the late 1970s, the neighborhood was not subject to any total repression and interference by state forces. Therefore, the struggles between leftists and rightists took a more spontaneous form, where the tactics people employed on the horizontal level, considering the power relations between different groups in the neighborhood, seemed to occupy a more extended place in the lives of the migrants when compared to the power relations on the vertical level. In this sense, due to the negligible role state forces played in the neighborhood during that period, even the shop owners, who played key roles in the production of space in the previous decade and who had power and political connections, seemed to deal with the repression of leftist militants by employing certain tactics of survival. The leftist militants were often said to visit the shops with the demand of *haraç* (protection money). While explaining their depression, *Sünni* shop owners revealed their sympathy to rightist militants and their distant approach to *Alevis* and leftist ideology:

> We experienced so many troubles from the 1970s to the 1980s. When you went out into the street, they interrogated you all the time by asking where you were from and your name. Our income declined enormously because we were put in a position of having to close the shop everyday... They wanted haraç (protection money) from us. They broke our glasses and windows.
>
> -Who did these things to you?
> -Rightists do not do such things.[340]

The hometown identities also had a political connotation during these years because sectarian identities could be discovered through hometown identities.

340 Extract from field research note no. 34, 3 July 2003.

As an example, if you were from certain provinces like Tunceli, or villages like Narlık of Çorum, then it was possible to identify you as *Alevi*. Everybody in the neighborhood had a knowledge about which villages and provinces were *Alevi* or *Sünni*. In this respect, the militants on the street, whether rightist or leftist, could interogate anybody to learn his hometown identity. This is why some individual shop owners from those provinces known for their support of ultra nationalist politics[341] mentioned the continuous pressure of leftist groups in this period in the following manner:

> In those years, being from Yozgat was accepted as a very bad thing. If you were from Yozgat, you were automatically considered as rightist, *Türkeşci* [supporter of Alparslan Türkeş, the leader of Nationalist Action Party]. It is not true at all. This divisiveness emerged after we had come here. Our village was very close to *Alevi* villages. We loved *Alevi* villages more than *Sünni* villages. After we migrated here, *Alevism*, *Sünnism*, being rightist, being leftist, all of them emerged. In the 1970s... It is a matter of people who come together from 40, 50 different places and settled and lived together. The politicians came and deceived our youths in those years. Beneath everything lied material reasons.[342]

Another shop owner from Yozgat, in similar style, explained his identification with the rightist ideology. He also described the fragmented nature of the public space in the neighborhood as follows:

> The most depressed period here for us were the first settlement years. The second most difficult phase was between the years 1978 and 1980. I witnessed the killing of a young person here in front of me. When someone got sick, we could not take them to the hospital out of fear. The space from the bridge up here [he refers to the part of the market place close to Samsun Highway far from the center of the market] was the forbidden zone. The

341 As mentioned in the introductory chapter, the provinces of Yozgat and Çankırı were known for their support of ultra-nationalist politics.

342 Material reasons here refer to the class dimension of the radical politicization of the neighborhood. This is often mentioned by respondents signifying the politicians' employment of economic deprivations of the youths to make them an integrated part of radical politics. In that respect, "the politicians deceived our youths" is a common statement made by many respondents referring to the late 1970s. Extract from field research note no. 34, 3 July 2003.

> other side of the bridge towards Üreğil was another forbidden zone. There were rightists mostly on the other side, but here the majority was leftists. We were under the rule of leftists. Being from Yozgat was so bad. They thought that people from Yozgat were rightists. But this was not the case. We also had many *Alevi* villages in Yozgat. We did not make a racial segregation.[343]

Despite the fact that these shop owners opposed their identification with rightist militants, they kept their distance from *Alevis* and leftist politics. This was made explicit in the course of the informal talks with them. These shop owners had developed certain tactics of survival in the insecure and threatening environment of the period. Since most of the illegal incidents and bombings in this period took place at night, almost all mentioned their preference of closing their shops early in the afternoon. According to these shop owners who felt most threatened in the period, to close shops early in the afternoon, not staying in the shop alone, unaccompanied by family members, and avoiding bringing guests or holding social gatherings in the shops can be counted among tactics of preventing direct attacks by the leftist militants. These tactics, in turn, contributed to serious loss of income in the market during the late 1970s. These protection tactics seemed to damage the natural flow of socio-spatial relations in the market place. The nature of shopping and the way shops were used seem to have a rather unique nature in Boğaziçi. Shops are, in general, social places where the relationship between customers and shop owners seems to be quite informal and to depend on the length of time they have known each other. In this sense, I never encountered a case where the shop owner stayed in the shop entirely alone during the day. The guests and informal gathering seem to be an essential part of daily life in the market place. The natural flow of life and consumption in the market place, in that sense, was blocked during that period. The survival tactics employed both by the inhabitants and the shop owners overshadowed the lively nature of the market. The restricted mobility of the neighborhood residents, in that sense, can also be counted among the reasons for the loss of income of the shop owners. According to the narratives of neighborhood residents, they had chosen to stay at home unless they were obliged to go outside during that period. The respondents reported the closure of many shops due to the financial and security problems during the period. The ones spatially close to the shop of "family X" felt this financial loss deeply. The son of such a shop owner describes the unhappy nature of his father at the time:

343 Extract from field research note no. 28, 17 June 2003.

> My father had treated everyone equally; he had not been interested in politics. Anyway, he told us how demoralized he had become in those years. He told us about a letter that he wrote to a carpet trader in Kütahya. The letter is something like that: "... here everything becomes so disordered. I cannot afford to pay you my debt. I cannot earn anything because I cannot sell anything. I will send you the goods I bought from you and I will close the shop". My father told us that at one time, rightists came for *haraç* at other times, came leftists.[344]

The shop owners suffered from serious economic impoverishment during those years. These shop owners were mostly the ones with rightist tendencies or the ones from the hometowns identified with ultra-nationalist ideologies as mentioned above. However, since the market place was the focal point of interaction in socio-spatial terms, most shop owners had suffered from continuous interrogation and pressure from young militants in these years, whether leftist or rightist. Sometimes, one's identity as a *Sünni* in a neighborhood dominated by rightists would not be sufficient for security. The shops as social spaces were described as being under continuous surveillance of militant groups and informants, therefore friends from the opposing sects or different hometowns might be taken as a proof of social proximity to the "other" group. During an interview with a shop owner from Kırşehir in Boğaziçi market, I encountered such an interesting story of a forced departure from Üreğil neighborhood to Boğaziçi. This shop owner[345] narrates how rightists accused him of being an informer for leftist groups. Since he had friends from both leftist and rightist groups, he was forced to leave his house and shop in the Üreğil neighborhood, the fortress of extreme right, and move to Boğaziçi neighborhood.

To sum up, the market place had become the focal point of interaction in the late 1960s and 1970s. Due to a number of socio-spatial reasons mentioned above, it witnessed the struggles between leftists and rightists. The heterogeneity of the neighborhood in terms of sectarian identities when combined with the exclusionary attitude of the powerful founders in the previous decade towards *Alevi* residents, led to the reaction of leftists towards these actors and their controlling the region by repressing *Sünni* majority in the market. However, as will be mentioned in the following pages, the control of the public space by the leftists provided little comfort in everyday life for the ordinary, politically inactive *Alevi* residents of the

344 Extract from field research note no. 14, 3 March 2003.

345 Extract form field research note no. 24, 14 May 2003.

region. The settlements of *Alevi* residents in a loose manner due to their latecomer status had put them into a disadvantageous position when compared to concentrated *Sünni* communities, who had formed a close containment of resistance against the opposing militant groups' intervening in their community regions.

5.5 Tactics of the Ordinary Inhabitants: Insiders' Space, Protection within the Community

Depending on the course of spatialization in the region, migrants from Kırşehir seemed to lose their power to a large extent in shaping the public space of the neighborhood in the 1970s when compared to initial settlement years, for a number of reasons. First of all, the migratory flows from different hometowns in the late 1960s and early 1970s had made the social composition of the neighborhood more heterogeneous. In that sense, migrants from Yozgat, Çorum, Tunceli and Sivas had emerged as densely populated communities compared to migrants from Kırşehir and Kırıkkale.

Secondly, due to their erroneous predictions related to settlement decisions, *Kırşehirlis* could not act as the founders of critical spaces in the neighborhood despite their status as early migrants. In this sense they could not own the lands that played a critical role in the development of the region.

Thirdly, since they had an established community identity from the beginning of the settlement period, they adopted a more inward looking and apolitical approach compared to the migrants from the less represented regions. In this sense, the shop owners playing a leading role in forming the neighborhood's political identity tended to be those from provinces other than Kırşehir. Despite their relative loss of capacity to take near-strategic decisions in the 1970s, *Kırşehirlis* greatly profited from the advantages of living together in a large homogenous community. Therefore, near-strategic acts in the initial settlement years at least provided a safe habitat for migrants from Kırşehir and Kırıkkale in the insecure environment of the time. Thus, early settlers who found opportunities to live with their *hemşehris* in spatial proximity felt powerful vis-à-vis leftist groups and the continuing struggles between parties in the public space and streets. Many people mention their secure position with reference to their identification with the towns of Kırşehir and Kırıkkale and their settlement in communities. Therefore, among these communities there was almost no mention of serious incidents or injuries among the community or near their residential areas in the highly political and violent space of the neighborhood. A woman from Kırşehir described the atmosphere in the neighborhood during the late 1970s in the following manner:

Everybody knows these people [leftist militants]. Both leftists and rightists know them. They killed each other at the train station because of the right-left conflict. These were the people we know very well. They were leftists, our people are rightists. It doesn't matter now. For example, I had friends from Sivas, they were killed in the events.
-How was your relationship with the *Alevis* residing here? Your neighbors?
-Majority were people from Kırşehir around here... Therefore they could not do anything to us... [*"Bize diş geçiremezler"* in migrant's words.] Most of Fahrikorutürk and Boğaziçi residents were from Kırşehir. Therefore, they could not go so far. But our neighborhood relationships were very good. We have grown together, played together...[346]

I encountered a reluctance to discuss *Sünni*'s attitude to *Alevis* in the narratives of *Sünni*, as in the above extract. On the one hand, due to the fact that they had been living together for years, they had *Alevi* friends and neighbors with whom they had good relations. On the other hand, particularly from the early 1970s with the politicization of the neighborhood, the distance between the politically loaded definitions of sectarian identities "us" and "them" had been widened.

Kırşehir community provided a safe social space for migrants even from other provinces. One of the *Sünni* neighbors from Yozgat living close to Kırşehir *hemşehri* community mentioned the secure environment of the region due the dense presence of *Kırşehirlis* in the region. As a survival tactic, families who had no obvious political standpoint preferred to live in isolation at home. In this sense, particularly during that period majority of the inhabitants narrate incidents refer to their loosening ties with public space. The male head of a family originally from Yozgat, describes their social isolation as follows:

We did not experience any difficulty during that period. I have six siblings; we all stayed at home and recited *zikir*.[347] While I was going to or coming from somewhere, nobody spoke to me. I went to work, to Çiftlik (*Atatürk Orman Çiftliği*) everyday. I arrived there at the time of morning call to prayers everyday [about 6 am in the morning]. However, one day, my son wanted

346 Extract from field research note no. 39, 15 Aug. 2003.

347 *Zikir* means repeating silently or aloud, the word Allah, the ninety-nine names of God, or prayers, which praise God.

to get off the train in Üreğil, but they did not let him. He was so scared. They asked him where he was from [his hometown]. He answered that he was from the right.[348] They were rightists too. They were *Türkeşçi*. "Nobody could harm you" they said to my son. Here people were all from Kırşehir, they [referring to the leftist militants] were scared of entering there.[349]

This protection within large communities seemed to bring some kind of safety perception to the narratives of community members. Despite the fact that Boğaziçi neighborhood and places close to it are described as the "fortresses of left", *Alevi* residents of Boğaziçi neighborhood describe their sufferings at the time being greater and more obviously when compared to *Sünni* residents. Particularly, the *Alevis* of Yozgat and Çorum living nearby or in the middle of the *Sünni hemşehri* communities describe their sufferings and insecurity more in relation to an area close to their residences. As gathered from the narratives, this insecurity seems to be mainly related to their tentative position between *Alevis* of other hometowns and their *Sünni hemşehris*. *Alevi* respondents make frequent references to the exclusionary and violent attitude of *Sünnis* while mentioning the period. In contrast, as will be mentioned in the following section, when the relations between *Sünni* and *Alevi* neighbors will be discussed, *Alevi* neighbors are described by *Sünni* respondents as being protective towards their *Sünni* neighbors whether they were their *hemşehris* or not during the fights between leftist and rightists. As mentioned above, some *Alevis* of central Anatolian provinces had chosen to settle in the neighborhood mainly taking the *hemşehri* relations rather than sectarian identity into consideration. Since *Alevis* were the latecomers, they had settled, in most cases, as a small community from the same village near a large community of a *Sünni* village of the same province. As an example, the migrants from an *Alevi* and *Sünni* village of Çorum settle nearby each other. In this sense, it is not possible to talk about a dense *Alevi* community settlement with the exception of the *Alevis* of Tunceli and Sivas.[350]

---

348 It is interesting to note here that hometown identities seemed to be often used as a means to get one's political standing in the incidents of the late 1970s as tackled from the narratives of the residents.

349 Extract from field research note no. 31, 26 June 2003.

350 As stated in the introductory chapter, some Central Anatolian provinces like Çorum, Yozgat and Sivas were known for their inhabiting *Alevi* villages in great number besides *Sünni* villages. Sivas is a Central Anatolian province known for the dominance of *Alevi* residents. In that sense the migrants from Sivas in *gecekondu* settlements are mainly *Alevi*.

Alevis of Tunceli and Sivas can be accepted as an exception since the number of *Alevis* from these settlements was exceptionally high and by settling in topographically the most unsuitable places to reside they were able to form themselves into a relatively autonomous community space. As understood from their statements, they were able to form a protected space. Outside of these protected boundaries, though, they might be treated brutally by rightists. Particularly the *Alevis* of Tunceli were perceived as the most radical and distant community according to the narrations of *Sünni* inhabitants. A migrant from Tunceli described the injury of his brother as follows:

> Yes, I remember those years... I was young then. We played outside in the evening. It was a winter day. I went to my uncle. We heard some gunshots. They shot my older brother while he was on his way to pick me up. The bullet went into his chest and went out from his back. Here these upper parts were all under the control of leftists' at the time. Enlightened people... In these lower parts, there are rightists [he shows the residential areas close to the market place]. In Üreğil there were extreme rightists.
> -Was he shot closer to Üreğil?
> -No, in these lower parts [he shows the residential areas of people from Kırıkkale]. There were some rightists groups here but here was constituted by leftists mostly.[351]

The *Sünnis* of Çorum and Yozgat put their *Alevi hemşehris* into a different category vis-à-vis the *Alevis* of Sivas and Tunceli. As will be mentioned in the following chapter, *Sünnis* of Yozgat and Çorum made references to their *hemşehris'* practicing some obligations of *Sünni* Islam in order to distinguish them from more radical *Alevis* of Sivas and Tunceli. Generally, the idiosyncratic way of *Alevi* practice of Islam is either neglected or ignored in *Sünni* residents' narrations.[352] *Alevi* respondents point out the

---

351 Extract from the field research note no. 40, 19 Aug. 2003.

352 In the discourses of *Sünni* migrants about *Alevism* and their neighbor's practicing of *Alevism*, it is inevitable to understand the tendency to identify *Alevism* as a version of *Sünni* Islam rather than a distinct practice. Therefore, the presence of mosques in *Alevi* villages, the *Alevi* inhabitants going to mosques for prayer or the *Alevis* who fasted during Ramadan are often given as examples to show the imposed sameness between *Alevism* and *Sünnism*. However, the construction of mosques in *Alevi* villages was mostly part of state policy favoring *Sünni* Islam over *Alevi* interpretation of Islam in the early 1980s. *Alevis'* accepting or practice of some *Sünni* Islamic obligations seems to be due to the social pressures of their *Sünni hemşehris* living close by.

main hostile activity against them by *Sünni* inhabitants was informing. *Sünni* neighbors had behaved as informants, frequently making their *Alevi* neighbors' identity explicit to the rightist militant groups. The narratives of a number of *Alevi* respondents reflect this informing activity by *Sünni* residents:

> -Did you experience any difficulty in the 1970s?
> -We were not able to say that we were *Alevis*. There were some men here, our neighbors [referring to his *Sünni* neighbors]. Their village was so close to ours, Uyurcak village... There were no means of transportation other than train then. At the train station they told others that we were *Alevis*. Ten, fifteen men with guns walked towards me intending to beat me. I said, "Are you not ashamed of yourselves? I am a civil servant. I am a member of *CHP*; I have nothing to do with other stuff. I am also from the village of Uyurcak". Then one of them said, "I am from the same village". They let me go with great difficulty.[353]

As mentioned in the above anecdote, to hide one's hometown identity could function as a tactic of survival for ordinary residents of the neighborhood against both leftist and rightist militants. Migrants were subject to continuous interrogations about their hometowns at the train station. Moreover, to live in a *gecekondu* neighborhood also made hiding identities difficult due to close encounters between communities. The train station was the most critical space for direct encounters between leftist and rightist militants. Almost all of the stories related to the atmosphere of the 1970s in the neighborhood include one important anecdote related to the train station. Migrants had no alternative but to travel by train for many reasons previously mentioned. The informal nature of the research setting had made the train as the only alternative means of transportation for a certain period of time. In this sense, particularly for men who were more mobile when compared to women in urban space, traveling by train was narrated as rather an intimidating experience due to close encounters with the militants. The train station was inevitably the meeting point of both Boğaziçi and Üreğil residents since it was the main and most critical point of social space which all residents needed to use. Consequently, there had emerged furious struggles between the leftists and rightists on this critical aspect of space in order to gain the control of the train station during the late 1970s, as described by the respondents who had witnessed these fights.

353 Extract from field research note no. 45, 1 May 2005.

The wife of a former leftist militant defined the extreme importance of the railway in this manner:

> Most of the incidents had taken place in the train. If the train at 7 am belonged to rightists, the other train after that belonged to the leftists. If someone had taken the wrong train by mistake, then, some bad events would definitely have taken place. Even slaughter could occur. Boğaziçi was mostly a *political* place[354] [she means space of leftists].[355]

In addition to the fact that the harsh struggles that had taken place, mainly at the train station, the struggles between leftists and rightists also occurred within the neighborhood mainly at nights as mentioned above. Due to dense settlement of *Sünni* residents in the neighborhood rightist militants collected haraç from the ordinary *Sünni* residents. The most interesting anecdote related to militants' way of persuading *Sünni* residents to pay haraç was narrated by a highly conservative *Sünni* shop owner. According to his report, the rightist militants carried out an attack to one of their *Sünni* neighbors' house at night and falsely claimed to the *Sünni* residents in the neighborhood that the *"communists"* were responsible. He described the incident as follows:

> There were more *Türkeşci* living close to our house. They shot the houses of their people at night then they told that communists had come and shot the houses. They collected money in return for protection against communists based on this lie. One day Hacı [his friend] kept watch at night to see who were shooting the houses. He saw that *Türkeşcis* came and shot their own people's houses. We got back some part of the money they had collected. They had spent some part of it. We gave this money to the mosque association.[356]

As Bruinessen (1996) mentions, religious extreme right fanned *Sünnis'* fear and hatred of the *Alevis* not only for the sake of collecting *haraç* but also

---

354 It is interesting to note that both *Alevi* and *Sünni* respondents use the concept "political" to refer to leftist politics. Political place, political people mainly signify the leftist politics in the discourses of the migrants. "Political" as a concept seems to have a connotation referring to the system opponent nature of leftist ideology.

355 Extract from field research note no. 46, 5 May 2005.

356 Extract from field research note no. 30, 25, June 2003.

for recruiting *Sünni* Muslims of the mixed regions. In order to do this, they spread "rumors" that *Alevis* had bombed a mosque or poisoned a water supply which unfailingly drew *Sünnis* into the extreme right camp.[357] Taking into consideration the highly chaotic nature of social relationships in the neighborhood, the tactics employed by each group towards the others determined the course of (social) spatialization in the neighborhood. The spontaneity of Boğaziçi and other *gecekondu* spaces emerged mainly from this fact during the period. The neighboring relations in mixed *gecekondu* settlements like Boğaziçi, within a context mentioned above necessitates close attention in order to understand the unique nature of these settlements, which will constitute the subject matter of next section.

5.6. Living Spatially Close to Each Other: "Sacredness" of Being Neighbors?

In general, being neighbors is considered a "sacred" part of everyday life in *gecekondu*. Being good neighbors, respecting each other's boundaries are the predominant values of *gecekondu* life. According to both research data and on my personal observations in different *gecekondu* neighborhoods, being a respected neighbor seems to be almost a norm in *gecekondu* life. It is important to note that the sensitiveness about the issue of "being a good neighbor" seems to be closely related to the competition over the ownership of land. Most of the time, the competition over space and the use and exchange value of the real estate constitute the primary concern in everyday lives of the migrants. As mentioned in the previous chapter, in the first settlement period, proximity to former members of one's own village was very important, since being part of a community was a precondition of defending their property against state forces. This mutual trust was necessary, not only to protect their house and land against state forces, but also respect each other's possesions. In this sense, the spontaneous and peculiar nature of land ownership and *gecekondu* construction patterns necessitated a relation of dependence. In a condition where the ownership to land could not be proven by legal documents, neighbors had to respect each other's boundaries in order to avoid arbitrary invasion. There are few cases where migrants described such disagreement with their neighbors over land issues, within the context of the most depressed period for them in the neighborhood. They gave frequent references to the importance and necessity of being good neighbors in *gecekondu* life in relation to these incidents. The sacredness of this behavior was also closely related to the fact

---

357 See Bruinessen (1996:8).

that the distinction between public and private spheres was more difficult to make in *gecekondu* neighborhoods when compared to middle and upper middle class districts of the city. Private lives of the families were often open to public view due to the particular aspects of *gecekondu* space and the domination of community life in these neighborhoods. In an environment where the boundaries of private and public space were indistinct, the migrants often mention the importance of having neighbors who were honest and helpful.

The utmost value given to neighbor relations in *gecekondu* life becomes noticeable at the point where the migrants narrate the continuing struggles between leftists and rightists in a confined residential area. In cases where *Sünni* and *Alevi* migrants were neighbors, unexpectedly, respondents mention their avoidance of open conflicts with each other. Particularly, some of the *Sünni* residents mention incidents where the neighbors from the opposite sect, despite their political position, have behaved in a very protective manner towards them for the sake of being good neighhbors in the highly conflict-ridden atmosphere of the period. There is almost no explicit reference to an open conflict between neighbors among respondents. *Alevis*, while referring to these years, use a more defensive narrative incorporating a threat perception and an explicit reference to their sufferings. The *Sünni* residents, on the other hand incorporate descriptions of positive and protective behaviors of their *Alevi* neighbors, as mentioned above. Despite their reaction against and distance from leftists and the *Alevi* identity in general, *Sünni* migrants adopt a mainly positive representation of their *Alevi* neighbors. A *Sünni* migrant from Kırıkkale illustrates the struggles taking place near their residential area by referring to their *Alevi* neighbors as follows:

> We blacked out the doors and windows with pillows during these years. In case they [leftist militants] had seen light and if they had guns; they might have opened fire on the houses. We are rightists you know. They [*Alevis*] did nothing to us since we were neighbors. My wife told me, a relative of hers, while he was about to get on his car one day, was surrounded by a group of leftists. My wife described how the leftists [their *Alevi* neighbors were among them] were about to beat him, she got out the house and told them that he was her relative. Then they let him go. They did nothing to you if you were a neighbor for the sake of avoiding dishonorable behavior.[358]

358 Extract from field research note no. 25, 15 May 2003.

Another *Sünni* migrant from Kırıkkale described how positive the attitudes of his *Alevi* neighbors have been towards him, in spite of having completely opposite political inclinations:

> If you behave well to your neighbors, they will be good to you too. For example, I have a neighbor from Sivas. I do not want to make discrimination, but they are *Alevi* and they were extremely leftist. They did much harm to many people ["*Tuttuklarını mahvederlerdi*"in migrant's words referring to terror years.] They knew that I am a rightist. But, Allah Razı Olsun, they have never mistreated me. Sometimes if they see me as sad, they come near and ask if I am all right.[359]

The importance and sacredness of neighborliness as a predominant value in *gecekondu* life, owing to peculiar socio-spatial qualities of these regions, seemed to dominate the radical political agenda of social spatialization in the 1970s.

One point of great interest to me as a researcher was that *Sünni* residents, while talking about the struggles between their leftist neighbors and rightists, often give spatial references to these neighbor's fights with the rightists on the opposite hill rather than those in their immediate residential area. This might reflect the real picture in these years because the conflicts would sometimes occur over long distances. Still, the inhabitant's inability to remember any conflicts in the immediate area may also give clues about their perceptions of the inappropriateness of aggression between neighbors. The similarity between the anecdotes on the same subject may explicate the common approach of *Sünni* migrants related to their *Alevi* neighbors in the following manner:

> Our left was occupied by people from Sivas and above us people from Tunceli were present. *They had worked actively during those years [she means their militant activities] but had done nothing to us. I do not know if this was related to the fact that we were neighbors. But they often fought with the people on the opposite hill. They fired at long range, from here to there* [by pointing the *gecekondus* on the opposite hill]. But neither the ones here nor residing there on the opposite hills did anything to us. They had fired weapons from these hills at each other. I had no problem with neighbors, my dear. One of them was from Tunceli. But they were so interested

359 Extract from field research note no. 25, 15 May 2003.

> in us. This was because they felt themselves as few in number. I cannot figure it out. They are just behind us, you see... *They could have done bad things to us, nothing like that had ever happened. They were always good to me.*[360] [Italics mine]

A rather conservative woman from Kırşehir who had mentioned her dislike of *Alevis* during the course of an informal conversation expressed a very similar attitude:

> There was a woman writing on the wall with mud. I talked to the commander of the gendarme station. I said to him "we are ashamed of her behavior but she does not". We erased it, and then next day they again wrote it down. On all of the houses there were writings [leftist slogans]. *But to be honest the same woman, they were from Sivas, had protected us against the attacks and therefore nobody had done anything to us. Since this woman's daughter was our neighbor, she was a friend of my daughter.* They knew us; therefore, they hadn't annoyed us. We all stayed at home at nights and blocked out all light. *We heard gunfire from the opposite hill.*[361] [Italics mine]

As in the anecdotes above, there was almost no mention of open conflict between *Alevi* and *Sünni* neighbors from the perspective of *Sünni* migrants. At worse, the relationship between neighbors of different hometowns had lost its intensity or totally broken as a result of the politicization of *Alevi* and *Sünni* identities. I only encountered two serious cases where a serious conflict between *Alevi* and *Sünni* neighbors had emerged. As a consequence of conflicts, *Sünni* and *Alevi* families exchanged houses in order to live among the migrants from the same sects, because if a small number of one sect resided in an area dominated by the opposite sect, they might have come under serious threat from the majority. I talked to one *Alevi* woman[362] from Kars who had taken part in such an exchange of houses with a *Sünni* family. She narrated how threatened they had felt during the period in Kayaş surrounded by a *Sünni* community from Çorum, despite her husband's distance from politics:

> *Çorumlus* got up in order to eat during Ramadan. They had stored gasoline in their backyards [I could not understand the meaning

360 Extract from field research note no. 36, 20 July 2003.

361 Extract from field research note no. 35, 18 July 2003.

362 Extract from field research note no. 56, 6 Oct. 2005.

> of storing gasoline at first. After asking a number of questions, I understood that she was scared that it would be used in possible attempts of *Sünni* neighbors to burn their house at night]. They fired their guns at nights. Since we did not have guns we felt very threatened. The ones [*Sünni Çorumlus* who exchanged their houses with the family] here went to Kayaş, to our house and we came here to their house. We had a very beautiful house there. I am very upset about that. Here the house was small. We paid the loan on the other house then we added rooms to the one here. Not only us but also some relatives of my husband were with us in Kayaş and they also came here after us. We exchanged houses.

They did this exchange without even taking their furniture. Some members of *Sünni* family from Çorum, the other party of the exchange clearly stated their feeling of discomfort because of having lived among an *Alevi* community, as narrated by their *Alevi* neighbors. An old *Alevi* woman narrated how they liked one of the two brothers in the family, but according to the woman, the other brother and his wife openly declared their feeling of uneasiness to their *Alevi* neighbors. She described the family's exclusionary approach towards them:

> We often visited them. However, one day, we heard that Osman told *Sünni* neighbors here that we were left among *Alevis,* and they made us feel insecure. In reality nothing bad happened. People from Yozgat and Sivas resided here. Ayşe the wife of Osman, behaved so well towards us, face-to-face... Then she talked behind our backs, she called us as *Kızılbaşlar* and unclean people. In those years it was said that they had some *hemşehris* in Kayaş and they told their *hemşehris* that they were not comfortable here.[363]

In general, during the course of the interview and according to my observations, other than the exchange mentioned above and the case of shop owner who moved from Üreğil to here, I encountered no case of such conflict and discomfort between neighbors from different sects. In the case that a few *Sünni* and *Alevi* families had been left among a community of migrants from the opposite sect, they might feel threatened and uncomfortable in the highly radical political atmosphere of the period. In general, the discourse of *Sünni* migrants related to their *Alevi* neighbors reflected their distance to and prejudices about *Alevism* in general, on the one hand, but on the other,

363 Extract from field research note no.45, 1 May 2005.

they made references to the incidents exemplifying the positive attitude of their *Alevi* neighbors towards them. However, the positive evaluations of the humanity and good will of *Alevi* neighbors, in particular did not evolve into a general optimism and tolerance free of prejudice towards *Alevi* society in the narratives of *Sünni* respondents. On the other hand, *Alevi* people more frequently spoke of their sufferings, and they were more guarded in their expressions regarding *Sünni* migrants compared to *Sünnis*. *Alevis'* good treatment of *Sünni* neighbors may not be necessarily linked to merely to good will. As narrated by *Alevi* respondents, *Sünni* neighbors often informed the leftist activities of their *Alevi* neighbors. Therefore *Alevis* were put into a position to develop certain tactics vis-à-vis their *Sünni* neighbors and treating their *Sünni* neighbors good was one of these tactics. *Sünni* respondents did, in some cases, mention their informing of the "unlawful" behaviors of their *Alevi* neighbors explicitly. The examples of this, I encountered mainly among the narratives of women since they were the important actors and the real arbiters of relations between neighbors. Their perception of the period is quite different than men's. While men seem to have knowledge about the incidents in the public space of the neighborhood transcending their immediate environment, women refer to their male children and their sufferings in the public space, and their neighboring relations during the period. They usually lacked a general knowledge about the incidents between groups in the wider public space of the neighborhood. But they did, however, give information about incidents, which were occurring around them. As an example, they may have a clear understanding of the uneasiness between neighbors but they were less well informed of events in the market place. Their betrayal by *Sünni* neighbors seems to constitute one of the important themes in *Alevi* women's narratives. As previously mentioned, *Alevis* and *Sünnis* who had migrated from different villages from the same province seemed to live as neighbor communities. An old *Alevi* woman described how his son was betrayed to rightists because of the fact that he pasted Atatürk's[364] picture on his notebook.[365] The one who had informed on him was the son of a *Sünni* neighbor from a village very close to theirs. For that reason, the child was surrounded by a group of *MHP*

364 "The reason behind most *Alevis'* positioning themselves in support of Republican ideology is related to the "laicist" reforms of Republican regime targeted *Sünni* communities that had been traditionally and historically the "other" of *Alevis*." See Okan, Murat (2004), *Türkiye'de Alevilik* (*Alevism* in Turkey), Ankara: İmge, p.93. Therefore, *Alevi* people identified themselves with Atatürk as a hero and Republican Regime's secular policies.

365 Extract from field research note no. 45, 1 May 2005.

sympathizers who demanded to see his notebook found it as described by the woman, and seriously assaulted him for being *Atatürkçü*, therefore communist. He was later found unconscious.

This informing was not unusual despite the fact that the neighbors were extremely careful about their attitudes towards each other in daily life. However, with reference to the period, the narratives of women point out a general estrangement in the social relations between *Alevi* and *Sünni* neighbors with the politicization of these identities. A somewhat expressive *Alevi* woman made the feeling of insecurity among *Alevis* vis-à-vis their *Sünni* neighbors explicit:

> -Have you experienced any uneasiness with your neighbors during the period?
> -Yes of course, a distance between us appeared. The girls were used to play outside in the garden together. We often went to visit our neighbors in the past, but during that period we didn't see each other that much. It was told that people from *MHP* put some machine guns in the garden of a neighbor from Yozgat [a *MHP* sympathizer]. But then, the youths of Sivas found and took them. It was said that their plan was to rake the entire neighborhood. You know here is an *Alevi* neighborhood. There was another story... People from Yozgat, because of the fact that there were so many *Alevis* in Boğaziçi had planned to fire all *Alevi* households. They left from Yozgat in buses but before they arrived here, the military intervention took place. They were left on the road.[366]

The loss of mutual trust on part of neighbors would be triggered in the coming decade when the state's policies of repression after the military intervention was enforced in the neighborhood through a survelliance strategy in which neighbors informed on one another.

5.7 Conclusion

The public space of Boğaziçi in the 1970s had witnessed struggles between leftist and rightist militants in close connection with the general political and economic context of the country. The spontaneity of *gecekondu* spaces in general and Boğaziçi in particular, emerged from the way these struggles had taken place in the socially mixed composition of the region and from peculiar social spatialization in the region. Due to the state's inability and

366 Extract from field research note no. 52, 10 Aug. 2005.

reluctance to intervene in the radically political space of Boğaziçi, events led to the spontaneous demarcation of neighborhood boundaries in a way that had no comparison at other times in the neighborhood history. The explicit struggles between militant groups or the implicit ones between ordinary residents through informing or giving covert support to militants exhibit a form of display of tactics on the horizontal level. Leftist militants applied pressure to once powerful groups who had been able to take near-strategic decisions in shaping public space in the previous decade. The exclusionary acts of once powerful groups in the 1960s had led to unforeseen spatial consequences that left the public space of the neighborhood dominated by leftists. The train station had constituted the main point in public space where extreme conflict had taken place. In the highly radical political atmosphere of the neighborhood, hometown identities had been radically politicized because they reflected different sectarian affiliations. At the same time the fact that different groups had lived together as neighbors being spatially close to each other a long period brought about a learning process between sectarian groups. Even though they appreciated their neighbors' good attitudes in special cases, *Alevi* and *Sünni* identities, in general, had become the predominant reference point in defining the "other" for both groups. However, it is not possible to neglect the heterogeneity among *Alevi* community related to hometown differences. Interneighbor relations had not taken an explicitly conflict-ridden form; however, these relations suffered great damage in the period, which would affect the course of social spatialization in the coming decade.

CHAPTER 6 - After 1980: Neoliberal Strategy, Identity Politics and the Politics of Space

6.1 Introduction

The military intervened in politics in 1980 with the claim to restore necessary conditions of democracy in the country. The strategy of the military regime was to restrain the radically political nature of public space and mainly urban terror. In conjunction with macro strategies of the military regime, the 1980s seem to be the years denoting the repressive interventions of security forces with the aim of avoiding active politics in public space of the locality. This strategic act of military regime to control and manage public space has led to a general degradation of active politics and aloofness in social relations. Not to speak about politics is defined as the only way to avoid possible troubles in migrants' expressions in such an atmosphere, dealt with in the following sections.

My research findings indicate that the strategic decision of the military regime to "depoliticize" the radically political space of the neighborhood was realized in three interrelated processes. First of all, at the local level, state security officials appear to have utilized a secret information network which involved encouraging some migrants to inform on other inhabitants in the neighborhood. Therefore, the public space of the neighborhood at the time can be best defined as a paranoid space where communication out of close *hemşehri* or relative cycle and political talks in public were avoided.

Secondly, at the macro level, the effect of neoliberal policies have been felt in the distribution of and claims to neighborhood space for a number of reasons, beginning in the 1980s. As mentioned in the second chapter, the legalization of *gecekondu* lands with the neoliberal policies of *ANAP* (Motherland Party-MP), the first civilian government after the intervention, increased the value of *gecekondu* land and led to the further commercialization of the real estate market in newly legalized "*gecekondu*" neighborhoods, as was the case for Boğaziçi. The reforms of decentralization in local government[367] system coupled with the neoliberal policies that triggered urban entrepreneurialism,

367 As stated in chapter two in detail, in three urban centers in Turkey- İstanbul, Ankara and İzmir- a two-tiered municipal system was established in 1983. District municipalities under metropolitan municipality began to serve as being responsible from the preparation and application of construction plans in their districts. By this law, most of peripheral *gecekondu* districts that had been built until that time were taken in the legal borders of district municipalities and connected to Metropolitan Municipality in administrative terms. See Heper (1987:1)

the low-income settlements were open to dynamics of urban land market. This brought unforeseen developments, in which the value of real estate in some low-income settlements collapsed, while some others gained value as a consequence of the contingent interplay of different dynamics such as the resistance of local groups to market mechanisms, the spatial location of the neighborhood and the strategic decisions of local government. The establishment of district municipalities under metropolitan municipalities in big cities inevitably made local political competition between political parties more responsive to the demands of "gecekodulu". In parallel to these developments, the public space of the research setting predominantly gained an economic meaning for its inhabitants, which in turn led to a general increase in the demands of the inhabitants for necessary services and the application of plans. This in turn, further increased the value of "*gecekondu*" houses and lands.

Thirdly, as part of the strategic act of suppressing active politics and dominating the ideological claims, particularly of the leftists, the military regime initiated and the succeeding civilian governments carried out a policy of privileging a *Sünni* interpretation of Islam as an antidote to the ideologies of Marxism and Kurdish separatism.[368] The *Sünni* version of Islam was presented implicitly as the formal understanding of religion by the state, and *Sünni* religious courses became compulsory during the period as a consequence of this policy. The state's repressive policies mainly towards *Alevi* leftists in the aftermath of the intervention, and their enforcement of compulsory religious courses in high schools alienated *Alevis* from the *ANAP* in the late 1980s, despite the fact that they had given support to the party in its initial coming to power.[369] As a result of these developments, the solidification of *Alevi* and *Sünni* identities with reference to their religious content seems to have determined the course of everyday life, beginning with the mid-1980s. The claims to space in economic and religious terms determined the form of political relations between the realm of strategy and tactic on the vertical level. The power relations on the vertical level seem to have become more reciprocal and the strategical realm more responsive to the tactical realm during this period when compared to other periods in the neighborhood history.

368 See Göner (2005:116) and also Okan (2004:93).

369 See Schüler (2002:168).

6.2 Paranoid Space: The Isolation of Social Relations in the Neighborhood

According to the narratives of the respondents, the first strategy of the state forces in suppressing the radically political space in the neighborhood seems to have depended on gathering information about "previously illegal" activities of the people, or of any kind of political talk in public, through migrants from the opposite sect or opposite political stand. The intimate relations between neighbors and *hemşehri* groups that had been greatly damaged in the late 1970s further receded due to a lack of confidence in social relations. The migrants' portrayal of the 1980s mainly embodies feelings of suspicion, anxiety and insecurity on part of those individuals that defined social relations. Speaking about politics was severely limited by the possibility of being reported by neighbors and friends to the police. The typical definition of the period can be exemplified by the report of a migrant from Kırıkkale in the following manner:

> After the revolution [he means the military intervention in 1980] there emerged a lot of trouble. Everyday they [the state security forces] took a trip here and collected people. Nobody could talk to each other. Informing activities often took place; therefore people hesitated to talk to each other. These were bad times.[370]

The village-like qualities of social relations was severely eroded during this period. Many people interviewed told incidents of being taken by the police for questioning. The highly repressive atmosphere of the early 1980s in terms of the power of state to define the course of social relations in an inflexible way, inevitably necessitated a distance from politics in the individual's perceptions of the period.

State suppression was used mainly against leftists in the aftermath of the military intervention. The active leftist militants of the 1970 suffered most from the pressures of the security forces because of the denunciation of their neighbors. The story of an active leftist militant, who is exceptionally, a *Sünni* himself, reflects this suffering. He imprisoned for about ten years as a result of the denunciation of his *Sünni* community. This was narrated by one of his closest *Alevi* friends who owns a shop in the market.[371] At some point in the conversation, his friend implied the unfaithfulness of his neighbors. In his words, "Even the ones whom he took to the school

370 Extract from the field research note no. 25, 15 May 2003.

371 Extract from field research note no. 19, 26 Apr. 2003.

in Üreğil[372] do not recognize and respect him now. He was also ostracized by his own people [referring to his *Sünni* neighbors]." I developed a friendly relationship with this militant during the course of the research. He shared details of his life story with me except for the incidents of the late 1970s. However, in his presence, his friends told anecdotes about how intimidating and powerful a figure he was for the rightist militants at the time. I also had the chance to observe the uneasy relationship he had with his *Sünni* neighbors. In spite of this he continued his friendship with a number of trusted *Alevi* friends and neighbors.

The *Sünni* and *Alevi* population, who had no direct relation with the political issues of the time, emphasized the apolitical nature of the public space in the 1980s. The question: "How were the tension-laden relations in the 1970s calmed down over time?" was mainly responded to by people with reference to the declining importance of politics during this time, and the increasing intensity of economic concerns of these individuals. This perception of the individuals about this period seems to have a close connection with the neoliberal policies of the late 1980s. At the macro level, during this period, the changing structure of the job market and the increasing unemployment rates resulted in the fragmentation of community based claims and solidarity. This made families and individuals more constrained by economic concerns, which contributed to the further detachment of their communal bonds. The common discourse in individual narratives portraying the two decades following the military intervention as a period, where political concerns were replaced by the struggle to make a living, seems to signify this process of individualization. A *Sünni* migrant defines this transformation in the everyday lives of the individuals as follows:

> After the intervention you could not even talk to your best friends unreservedly. Now everything is good. People understand that there is no good coming from politics. Everybody is after earning their daily bread. Life was hard for everybody at the time. These issues had lost their importance; there are no left or right anymore.[373]

---

372 Due to the fact that Üreğil neighborhood was occupied by mainly rightist militants, it was very difficult for people from Boğaziçi to pass there. However, children should go to the school on the other side of the bridge in Üreğil. This militant had taken children and accompanied them to the school.

373 Extract from field research note no. 23, 5 May 2003.

Those who had actively participated in the political struggles in the late 1970s, along with their families, emphasized the very same premise, that both leftists and rightists were the victims of the period. A member of the previously militant leftist family from Tunceli narrated the nature of the social relations in the 1980s as follows:

> There was no longer a tension like that; even the leftists and rightists had become friends with each other. Another reason was they had lived together for so long. Everybody knows each other here. Now the leftists and rightists sit together, talk and discuss politics. In addition, the events and relations took a more economic turn. To earn bread money has become important; people do not care about left and right, any longer... They believe in who gave them more money.[374]

As mentioned above, and will be discussed in detail in the following sections, this change of approach among migrants towards political issues and their emphasis on economic concerns had a close relationship with the preeminence of economic determinants in the social spatialization of the locality by means of neoliberal policies.

6.3 The Effect of Neoliberal Policies on Neighborhood Space

The uneasiness in social relations because of the lack of trust following the period of open struggles between leftists and rightists, and the repression of active politics in the public space should not lead one to a quick conclusion that the public space had become, to a large extent, depoliticized in the 1980s. On the contrary, the legalization of *gecekondu* neighborhoods through the neoliberal policies of the 1980s had made politics more embedded in the production of public space when compared to previous periods. This had one strong relation with the fact that the relations in the production of space had come to be determined more by economic motives and concerns by the legalization of most *gecekondu* neighborhoods, beginning from the 1984 onwards. With the *gecekondu* law established in 1984, most of the *gecekondu* settlements that had been built before that time were legalized. By means of legalization in these neighborhoods, the estate prices and rents in *gecekondu* districts increased and these spaces became more open to the influence and dynamics of urban entrepreneurialism. Particularly, the early settlers benefited from this condition. Boğaziçi was also taken within the

---

374 Extract from field research note no. 40, 19 Aug. 2003.

framework of legalization activity by the introduction of the first extensive construction plan of the region. It began to be prepared in 1985 during *ANAP* district mayoralty. Law no. 3194 gave the right to authorize construction plans to the municipal governments in 1985. The Mamak Municipality was established as a district municipality, an extension of Ankara Metropolitan Municipality and was given the authority to make and apply construction plans. The *gecekondu* settlements in the periphery of Ankara were joined with the city as legal suburbs under the municipal government during this period.

Up until 1989, approximately 10% of the Mamak region had been included in the framework of the construction plan and legalized according to municipal policy. The process of legalization in only a limited area of Mamak in the late 1980s is said to reflect the inability of the first district mayor from *ANAP*, to prepare a well-defined construction plan and operate it efficiently. As supported by the reports of some inhabitants and municipal officers, the mayor gave priority in the legalization process to some neighborhoods from which the party had received the main support. In this sense, those parts in migrants' narratives concerning the political acts of the first district mayor, reflect the beginning of feelings of favoritism and injustice in local politics. The major legalization activity in the region took place between the years 1989 and 1995. Social democrat mayors were in power during this period and most of the Mamak *gecekondu* settlement was legalized. The remaining area was not legalized due to its geological inconvenience for construction. Many respondents defined Selahattin Öcal, who came to duty in 1989, as the most effective social democrat mayor by both *Sünni* and *Alevi* residents. However, some *Sünni* respondents also signified the decisions that were implemented during his term favored mainly *Alevi* migrants. In this respect, the aftermath of the highly political atmosphere of the late 1970s seems to take place in the memories of the migrants with reference to the struggle to get legal rights with the best possible conditions to their lands and houses vis-à-vis one's neighbors, and "other" groups, and in relation to this, by an awareness of the possibility of biased treatment on the part of district municipality according to sectarian identity. The increasing knowledge of migrants about local politics, their awareness of nepotistic policies and their willingness to utilize favoritism for their interests in relation to real estate had increased the power of particularly early comer migrants. This implied that they were no longer transitory migrants but active participants having a say in local politics.

The most remarkable aspect of local politics seems to have been constituted around the competition mainly between *Alevi* and *Sünni* communities over legalizing and planning the neighborhood space. This reflected a significant transition from an explicit and harsh political struggle in the public space, namely a space of politics, to an implicit struggle for space, that is a politics of space. The reciprocity and responsiveness of the relations between the strategical and tactical realm, on the vertical level, intensified with the increasing power of migrants to define the fate of local politics. This is the reason why the respondents narrated the activities of the social democrat mayors, Selahattin Öcal and İsmail Değerli respectively, between 1989 and 1999, within the sphere of favoritism. In particular, *Sünni* migrants defined the period as advantageous for the *Alevi* population. One of the municipal officers who was actively engaged in the legalization process at the time and who is an *Alevi* himself admitted these incidents of favoritist practices as follows:

> The private construction offices under oath determined the condition of lands at the time. 30% of people's land was considered as the right of the municipality and people were obliged to pay this amount to the municipality in order to be included within the context of the construction plan. However, the holder of the right decided the value of the land. The municipality was the injured party because of that. For example, if they fixed the value of their land as 10, the municipality could not say it was 5. We constructed committees and determined the jointly owned lands. The offices under oath did the ground measurements. We distributed the technicians to six different regions. We, in all rounds, sent officers to different regions in order to avoid favoritism. But things like did happen, we could not prevent them.[375]

Apart from the nepotistic policies of the municipality, the legalization process itself had loopholes that were open to exploitation. As mentioned in the second chapter, the responsibility that was given to the private construction offices under oath, and the informal relations between the inhabitants and officials caused migrants to influence these relations to their advantage. The narratives of the Boğaziçi inhabitants reveal the fact that the legalization of the houses during the period did not lead to an automatic increase in the prosperity of all. The informal moments in the actualization of legalization brought injustices to different groups in society depending on their political and sectarian standing. The ones having legal title deeds to their land were put in a more disadvantaged position with legalization vis-à-vis the owners of the *gecekondu* houses on state owned lands. In other words, the ones who legally bought the vineyards of the Üreğil villagers

375 Extract from field research note no. 42, 21 Dec. 2003.

could not take advantage of legalization much as those who had built their houses on state owned lands because the migrants who had their houses on state lands got their formal title deeds according to the declaration they gave about the size of their lands to private construction offices under oath. The inhabitants often gave examples of the exploitation of the relationship between the inhabitants and officers under oath. As a tactic, those who had no title deeds gained the right to the state owned lands on which they had built *gecekondu* houses depending on their verbal declaration about the size of the land. They used the opportunity to declare their land sizes larger than the normal size of their gardens and extended their rights to the land by including the surrounding state lands within their declaration. The poorest migrants who settled on the steepest parts of the neighborhood suffered from the legalization process the most. These places were included in the construction plan as green areas, which were to be transformed into public recreational areas in time.

In addition to legalization, the integration of neoliberal consumption spaces [the opening up of different shops selling international brands] increased the value of real estate prices in the market place. Particularly, the established shop owners regained their power in economic and political terms, beginning by the late 1980s. In general, they benefited from the application of the construction plan in the region and the transformation of public space. This process retransformed them into powerful agents who could take near-strategic decisions. This restoring of power on the part of shop owners contributed to the spontaneity of neighborhood space in terms of posing an alternative spatial development to the mainstream tendency of apartmentalization in *gecekondu* areas.

Apartmentalization of *gecekondu* districts is conventionally considered as the integration of these districts with city space and the solution of the problems, at least in appearance, in the public perception. However, apartmentalization of *gecekondu* districts, in some cases, offered no real solution to the problems of low-income settlements; on the contrary, it triggered further problems with its unplanned nature and its potential to alter the social rhythm of everyday life in these districts. Moreover, the legalization of the neighborhoods did not, in reality, inevitably bring apartmentalization, as expected by the modernization approach. As mentioned by Brenner and Theodore[376], the neoliberal capitalist territorial organization is always unpredictable and deeply contested because, free from the market value of a certain land, social attachments to place persist as people struggle to defend the everyday practices and institutional compromises from which

376 Brenner and Theodore (2004:8).

capital has sought to extricate itself. Local actors may resist the neoliberal act of "creative destruction"[377], not only to defend everyday practices or as a result of their emotional attachment to *gecekondu* space but also for the fact that durability of the built environment as it is can be more beneficial for its holders. This potential spontaneity in the neoliberal strategic spatial practices operates similarly as a factor behind the late apartmentalization in Boğaziçi. This subject will be dealt with in detail in the following sections.

It is worth mentioning that, with the operation of neoliberal policies, the economy had come to constitute the main axis of everyday life. The public space was used and produced by the social relations that were predominantly determined by the economic relations and assets of individuals. The proliferation of consumption choices has led to the diffusion of new shops selling national and international brands in the market place and the modern means of consumption into the economic transaction process. The presence of the typical traditional *gecekondu* shop, together with chain supermarkets that are preferred to a large extent by the migrants for food shopping, reflected such a hybrid structure. The contrasting view of a collapsing old *gecekondu bakkal* alongside recent advertisements of newly emerged goods and a computer print of the statement "visa card is available" hung on its window, is a commonly encountered scene in the recent outlook of the market. The social spatialization in the neighborhood, and particularly in the market place, was increasingly determined by economic relations in the 1990s. The value increases in *gecekondu* land, along with the neoliberal policies mentioned above and the changing structure of the local government system, increased the effect of favoritism and clientalism in local politics; this will constitute the subject matter of the following section.

---

377 This concept is developed by Brenner and Theodore (2002:15). According to the authors, the actually existing neoliberalism should be analyzed with reference to two dialectically intertwined but analytically distinct moments: "the (partial) destruction of extant institutional arrangements and political compromises through market-oriented reform initiatives; and the (tendential) creation of a new infrastructure for market-oriented economic growth, commodification, and the rule of capital. Two important caveats must be immediately added to clarify this conceptualization of actually existing neoliberalism as a process of institutional creative destruction." Ibid., p.15.

6.4 Favoritism as the Side Effect of Neoliberal Strategy: The Increasing Importance of *Alevism* and *Sünnism* as Part of Politics of Space

The increasing effect of favoritism towards ethnic and sectarian identities in local politics appears to be a reality of the 1980s. Schüler[378] attributes this to two main reasons; the decreasing charisma of political leaders in conjunction with the loss of ideological cores in party programs, and the increasing domination of identity politics along with the declining validity of class-based politics following the collapse of the socialist bloc. The repressive strategies of the military regime, mainly towards leftist ideology, and its prioritization of *Sünni* Islam as the official religion of the state led both to the inability of the social democrat parties to use ideological means for political support in the aftermath of the intervention and to an attempt on the part of the *Alevis* to revitalize *Alevism* as a belief system. In parallel to these developments, the social democrat tradition in local politics followed a more clientalistic way of sustaining support from the urban poor rather than through employing ideological discourses.[379]

As was also the case in Boğaziçi, in the late 1980s and early 1990s, the domination of social democrat mayors in local politics in largely low-income neighborhoods for a brief time put the *Alevi* communities in a particularly advantageous position of using their political connections in the appropriation of space vis-à-vis other groups. The *Alevis'* support for the SHP (Social Democratic People's Party) appears mostly to be a consequence of the party's favoring an individualistic understanding of religion. The SHP explicitly stated its position in debates about Quran courses in 1986 and 1987, and raised the issue of the impossibility of gaining any concession from traditional Kemalist and laicist understanding.[380] The intimacy between the SHP and the *Alevi* society is exemplified by SHP's nomination of *Alevi* candidates in the 1987 national elections, particularly in cities where the social polarization between *Alevis* and *Sünnis* was rife and the support of *Alevis* for the SHP was high.[381] As mentioned in previous sections, the local government system in big cities was reformed with the introduction of district municipalities alongside metropolitan municipalities. Social democrat mayors had been in power in the late 1980s in both Ankara Metropolitan Municipality and Mamak District Municipality. This strengthened the power of social democrat officials to employ clientalistic

378 See Schüler (2002:117).

379 Ibid., p.119.

380 Ibid., p.171.

381 Ibid., p.184.

means to gain the support of the *Alevis* in the region. From the perspective of the *Alevis*, the 1980s in local politics seem to have been beneficial for their claims to space in economic and political terms. If their disadvantageous position as latecomers to the city is taken into account, their need to make greater use of the clientalistic means as compared to *Sünni* migrants seems to be understandable. As also shown by the findings of *gecekondu* research by Gökçe et. al[382] *Alevis* mainly answered the question: "How do you resolve your problems in state institutions?" as through finding acquaintances in state departments rather than by themselves individually. Clientalism for *Alevis* appeared to function as a tactical means to overcome the obstacles put forward by the strategical realm.

In this respect, in the aftermath of the highly radical and fierce battles of the late 1970s, the reality of the last two decades in the neighborhood has been an implicit competition that is dependent on the use and production of public space between different groups, mainly between the old rivals, the *Alevis* and *Sünnis*. Therefore, as mentioned in the previous section, favoritism, that has become embedded in local politics beginning from the 1980s, has replaced the explicit competition between *Alevis* and *Sünnis* in the context of leftist-rightist struggle. In this sense, the municipal treatment of the residents and the production of public space in Boğaziçi have acquired a highly political content, depending on the political and sectarian affiliation of the mayor. The interviewees narrated stories of the municipal service provision and the financial penalties given, particularly, to shop owners with direct references to *Alevi* and *Sünni* identities or distinctions. *Muhtars*, as the local headmen, are the most important actors whose role is to articulate the local needs, demands and knowledge with the municipal level politics. As shown in the findings of Erder's[383] research into one of the biggest low-income settlements of İstanbul, migrants counted *muhtarlık* as the most visited local institution, before mosque, municipality, coffeehouse, and political parties. It would appear that *muhtars* were the preeminent witnesses to and actors of the politics of space in the region by the 1980s, as mentioned above. One of the previous *muhtars* of the neighborhood gave an example of the favoritist policies of a social democrat mayor of Mamak in this way:

> Mamak Municipality was social democrat at the time. There weren't any people from the right in the municipality then. When Selahattin Öcal came into power, all department heads were

382 Gökçe et. al (1993:311).

383 Erder (1996:272).

> *Alevi*. Services were given to places where *Alevis* were densely residing. To give an example, we had a legally appropriated site for a public bazaar on the upper side in our neighborhood when İsmail Değerli [the mayor succeeding Selahattin Öcal] came into power. However, he had the bazaar constructed in another place [he refers to the place on the boundary between the Boğaziçi and Akşemsettin neighborhoods where *Alevi* people were densely residing]. I mean he did not allow the bazaar to be built on the suitable place left for that purpose; he expropriated another place and had the bazaar built there, close to his own people.[384]

The same person told a story where the naming of a school had become a problem between him and the *muhtar* of an adjacent neighborhood. He explained how he had applied to the Ministry of Education for the erection of a primary school in the neighborhood. According to this narration, the construction plan, including the plan of the school, was accepted in 1989. However, the *muhtar* of Akşemsettin, an *Alevi* woman, insisted on naming the school "*Devrim*" [meaning "revolution" in Turkish, the motto of the left in the late 1970s]. According to the *muhtar*, this was a totally illegitimate claim on the part of the *Alevi muhtar* since he initiated the construction of the school in his neighborhood. He explained the effort he had made in naming the school "*Şehit*[378] *Murat Eroğlu İlkokulu*" in the following manner:

> She used her connections a lot and tried to give the name "*Devrim*" to the school. I went to the *kaymakam* and the Ministry of Education and suggested the name of a soldier who had lived in our neighborhood and died while serving in the army. The ones living in Akşemsettin were using their votes in that school. *Alevis* had a majority in the Construction Council. The *muhtar* had printed voter cards under the name of *Devrim İlkokulu*. For that election, the school was admitted as *Devrim İlkokulu*. Then we gave it its formal name.[386]

Similar to the complaints of the *Sünni muhtar*, *Sünni* shop owners define the period in the late 1980s and early 1990s by giving references to the biased attitudes of *zabıtas* (municipal police officers) working under the

384 Extract from field research note no. 29, 20 June 2003.

385 "*Şehit*" refers to a person who dies while serving Turkish state or Islam. The rightists name places after şehits so as to signify their nationalist sensitivity.

386 Extract from field research note no. 29, 20 June 2003.

social democrat mayors and who are defined by shop owners as mostly *Alevi*. They point out the main reason behind the close surveillance of social democrat *zabıtas* during the period as the identification of the market space with *Kırşehirlis*, therefore considering it as the space of *Sünni* and religiously conservative people. The social democrat municipal attitude in the late 1980s and early 1990s regarding this particular space is narrated as always more discomforting vis-à-vis the attitude of the recent religiously-oriented conservative municipal cadre. Almost all of the shop owners mentioned their sufferings during the government of the social democrat mayors. The most often narrated complaint was the frequent visits of *zabıtas* to enforce and supervise the acts of the shop owners in compliance with the rules and regulations. Financial penalties for the proper and healthy functioning of the market are given as the main instruments of politics, according to the traders. Both *Sünni* and *Alevi* shop owners mentioned the feeling of pressure as a result of the continuous control of the municipal officers during that period. *Alevi* shop owners narrated their sufferings, like the *Sünni* shop owners, due to the recognition of market as a conservative space by the 1980s on. One of these *Alevi* shop owners compared this stressful period with the actions of the religiously oriented municipality of today as follows:

> When the leftist mayors were on duty, inspectors [*zabıtas*] were coming from the municipality every two weeks. Now these ones [recent municipal police officers from *AKP*] come once a year and they, most of the time, return without fining anybody. These people do not touch the poor and the needy ones. If I wasn't aware of their real motives [referring to the populist and religiously fundemantalist motives of the party to collect votes], I would regard their services as far better than the leftist municipalities.[387]

Sünni shop owners mention the relative depression they felt during the period when the social democrats were in charge of the Ankara Metropolitan and Mamak district in contrast to the quite peaceful and profitable period that came in the aftermath of the military intervention of 1980. They described the period of the social democrat mayors as one of the most stressful periods, despite the fact that they also narrated stories related to

---

387 Extract from field research note no. 7, 5 Nov. 2002. In spite of the fact that *AKP* party was not in power in that period, some migrants, in their recent expressions used *AKP* in order to refer religiously oriented parties with reference to past politics also.

their power and political connections against the acts of municipal officers. Also on this issue, respondents signified "family X" as the leaders of local resistance against the "unjust treatment" of municipal officers. The family, by using their political connections and the local social power they had, appeared to lead the resistance on the part of the shop owners. According to the narratives of the shop owners, during the term of İsmail Değerli, a social democrat mayor, *zabıtas* prohibited the exhibition of goods in front of the shops on the street. Shop owners were obliged to put their goods inside their shops.The "family X" had a grocery and for them particularly, the exhibition of vegetables and fruits in front of the shop was a necessity. A member of the family narrates how they found an acquaintance who held a critical position in the municipality and how they resisted the control of the *zabıtas*. He particularly emphasized the fact that this constituted a model for other shop owners and the supervisions become rare as a result of this act. Some other shop owners' narratives, exemplifying the leading and powerful shop owners in the region, support this narrative as follows:

> -Who would play the leading role in transmitting the complaints of shop owners to local politicians?
> -The big shop owners, the leading ones... For example, in the past there was a municipal police officer... At the time of the leftist municipality... The Karayalçın period... This person who groused about the shop owner was *Alevi* but the grocery owner was from *MHP*... They were from opposite factions. In order to force him to move back two boxes [from the pedestrian way], he often fined him. He said: "Take the boxes in, do not put them outside." We could not even leave a chair outside at the time. They [the leading shop owners] come together among themselves, may be 15 shop owners. They did a sudden visit to the municipality and also arranged for people from important places to be there. They forced the mayor to warn the principal [of the municipal police officers]. The principal was persuaded, then the issue was concluded.[388]

The power of the shop owners was further restored with the declining power of the social democrat dominance in local politics. Political Islam gained momentum, starting with the noticeable victory of the pro-Islamist Welfare Party (Refah Partisi-*RP*) in the 1994 local elections. This was also the case in the Mamak and Boğaziçi neighborhoods.

---

388 Extract from the field research note no. 14, 3Apr. 2003.

The 1994 local elections signified the beginning of a new period in the history of low-income settlements. *Gecekondu* residents, who had been defined as the "other" in Turkish politics up until the 1994 local elections, accomplished a type of democratic revolution against the central elitist powers, and this was the first time where those people defined as "the other" for years had direct power in determining urban policies.[389] Some relate the success of the *RP* to its appeal to the economic needs of low-income settlers rather than its religious orientation.[390] Besides the economic factors, the local political success of the party depended mainly on the migrants' disappointment with the inefficiency of the local services provided by the previous social democrat parties, the suppression of leftist politics in the peripheral settlements by state security forces from the late 1970s, and the success of the *RP* in grassroots organization and campaigning. The disappointment felt for years by the residents of the peripheral settlements towards the social democratic tradition, once a strong basis in the peripheral settlements, was clearly revealed in the 1994 and 1999 local elections. The *RP* aimed at mobilizing those urban poor who had suffered from the liberalization policies of the 1980s which had had such a negative impact on the peripheral social and economic groups.[391] It attracted 19% of the votes nationally and gained 329 mayoralties, including Istanbul and Ankara. The victory of *Adalet and Kalkınma Partisi- AKP*, (The Justice and Development Party) with a strong Islamic basis who came to power in the 2002 national elections, was announced by a widely read newspaper, Sabah, just after the Election Day as an "Anatolian Invasion".[392] These developments triggered the divide between secularist and Islamist groups, and from the perspective of the urban elite, the return to all kinds of community formation in low-income neighborhoods is considered dangerous for national unity, as mentioned in chapter two.

With the popularization of conservative Islamic practices particularly among *Sünni* population living in low-income settlements, the dominance of religiously oriented parties in local politics opened a new phase in urban politics. In parallel to these developments, shop owners mentioned the relative comfort they experienced since Melih Gökçek came into power.

389 See Yalçıntan, C. Murat & Erbaş, A. Erdem (2004), "Mekan ve Siyaset", *Birikim*, 179, p.40.

390 See Özler, İlgü (2000), "Politics of the Gecekondu in Turkey: The Political Choices of Urban Squatters in National Elections", *Turkish Studies*, 1(2), p.53.

391 Dağı, İhsan (2005), "Transformatin of Islamic Political Identity in Turkey: Rethinking the West and Westernization", *Turkish Studies*, 6(1), p.25.

392 Demirtaş and Şen (2007:97).

As mainly reported by *Sünni* shop owners, the control and supervision of police officers became looser, due to this transformation in local politics. In other words, shop owners related the relaxation of municipal control beginning by the mid-1990's to the identification of the market place with *Sünni* conservatism. Many shop owners emphasized this connection in a manner similar to the anecdote below:

> We open our shops on Sundays. The head of the municipal police here in Mamak is from *AKP*. In my opinion, they think that all shop owners here are *AKP* supporters. Therefore he does not pay any visits to the region. We open the shops on Sundays and we do whatever we want.[393]

As mentioned above after the 1980s, the shop owners, particularly those who could adapt their shops to the changing consumption patterns or who opened new shops in response to the newly emerging demands, gained certain economic and political power as a result of the neoliberal policies. These shop owners were those who had settled the market earlier and usually they had more than one shop in the region. Despite their complaints about the period of the social democrat mayors, the leading ones particularly had the power and connections to resist the acts of the municipal officers of the time. In the late 1990s, regardless of the effects of the economic crisis, the established shop owners primarily seemed to gain increasing power in my observations. These are the ones who seem to have the power to make near-strategical decisions that might unintentionally shape the neighborhood space. This will constitute the subject matter of the following section.

6.5 The Near-Strategical Decisions of Shop Owners vis-à-vis the Paradoxes of Strategical Realm

Unlike the estimates of the neoliberal strategy, the legalization of *gecekondu* houses and lands by means of the measurements of private construction offices under the supervision of a two-tiered municipal system had not necessarily attracted large construction firms willing to invest in the peripheral lands on which low-income settlements had already been established. On the contrary, as Erder[394] mentioned, the preparation of construction plans and their application brings a settlement to all social layers and power structures within *gecekondu* neighborhoods as they stand. Since planning attempts

393 Extract from field research note no. 5, 10 Oct. 2002.

394 Erder (1996:158).

and applications tend to protect the existing structure, this conserves the legal small parcels as they are which, in turn, avoids the entrance of the big capital-rich construction firms to such low-income neighborhoods. The real estate system in these neighborhoods appears to depend on small property ownership and individual legal title deeds to land and houses. Therefore, as elegantly pointed out by Erder[395], the new physical and social relations emerging after legalization in such neighborhoods, hinder the two groups on the extreme poles, the poor and the big capital, from entering the real estate market in these regions.

The neoliberal shaping of urban space brings unintended and contingent happenings.[396] As Weber[397] discusses in her article on neoliberalism and urban redevelopment, the powerful capital will only invest in urban property if it is recognizable beyond its unique character embedded

---

395 Ibid., p.158.

396 Even there emerges a recent theoretical distinction/classification between "ideal-type" versus "contingent" neoliberalism. Ideal type of neoliberalism depends predominantly on Hayek's and Friedman's selective reading of classical liberal project in opposition to egalitarian turn within liberalism that began with John Stuart Mill and continued through the work and influence of Keynes. See Hacksworth, Jack and Moriah, Abigail (2006), "Neoliberalism and Contingency and Urban Policy: The Case of Social Housing in Ontario", *International Journal of Urban and Regional Research,* 30(3), p.511. The core principles of ideal type liberalism can be listed as follows: 1) The individual is the normative center of a society; 2) The market is the most effective means through which individuals can maximize their own utility functions; 3) State actions that interfere with either individual autonomy or market relations lead to an autocratic society regardless of their intentions (İbid., p.511). "Contingent neoliberalism" conceptualization emerges, on the other hand, from a three groups of criticism to the contested nature of ideal-type definition of neoliberalism as defined by Hacksworth and Moriah (2006:512). The first group has challenged the ideal-type neoliberalism on geographical grounds. "They argue that localities filter and define neoliberalism at least as much they (localities) are filtered and defined by such ideas." A second group in close relation with the first group has challenged the internal consistency of neoliberalism itself. They suggested that neoliberalism entails a form of state intervention that contradicts one of its basic premises that of minimalist state. A third group of thought challenges the emphasis on neoliberalism within cities given the presence of other parallel processes and ideologies, such as neocorporatism, neocommunitarianism and neostatism.

397 Weber, Rachel (2002), "Extracting Value from the City: Neoliberalism and Urban Redevelopment", in Brenner, N. and Theodore, N. (eds.), *Spaces of Neoliberalism,* Blackwell Publishing, p.183.

in space and if it is believed to sustain short-term returns. According to this author, if these conditions are not met, disinvestments may occur in certain regions, constituting the reason behind the uneven urban spatial development within the context of redevelopment of urban land under neoliberal strategy. This argument may shed light on the reluctance of big investors to invest on some *gecekondu* neighborhoods. The complex nature of property ownership patterns in *gecekondu* neighborhoods might contribute to understanding why the big investors were reluctant to invest in *gecekondu* settlements despite the legalization of the neighborhood. If the majority of the neighborhood residents, under the leadership of influential actors, act in total compliance with *muhtars* and local politicians to attract such investment, the neighborhoods may become middle and upper middle class suburban districts. Even in such cases, there is a great possibility of serious problems arising due to under representation of certain groups.

The legalization of the neighborhood, and the land value increases have not brought consequential apartmentalization in Boğaziçi, partly due to the reasons but mainly because of the increasing power of influential shop owners in taking near-strategical decisions. The dependency of apartment construction on the decisions of the shop owners in the region had one important spatial cause. Spontaneous apartmentalization in previously *gecekondu* neighborhoods should for practical reasons be initiated beginning from either side of the main roads. Only then, for practical reasons can apartments in other parts of the neighborhood be subsequently built. The constructors are also more willing to build apartment houses principally adjacent to the main roads. This common pattern of apartmentalization is also valid for Boğaziçi. As mentioned above, shop owners owned the lands along the main road in the neighborhood passing through the market. Depending on the narratives of the respondents, some shop owners, for reasons of self interest, have acted near-strategically and resisted the requirements of the construction plan that had changed for a number of times. In their narratives, shop owners; make frequent references to their own or other shop owner's acts of altering the course of spatial development in two main ways.

First of all, some are recorded as saying they slowed down the application of construction plan by using their connections with municipal officers. The construction plan is intended to give the main road its final form. The construction of the road according to the plan is not possible without the partial demolition of some shops or houses on the edges of the main road, which would work to the disadvantage of some shop owners.

Secondly, as stated by the respondents, some shop owners who are also owners of land may not prefer to sell their lands to the constructor to allow apartments to be built in place of their shops and houses due to the unprofitable nature of this transaction for them. If they hold sufficient financial assets, they mention their preference to build their own apartments instead of using constructors. The already existing apartment buildings in the region had been built in this way. If they chose to sell their lands to the constructor, what they received in return would be one or two flats from the constructed apartments. Because of this, in contrast to the expectation of modernization approach that the value increase in land and planning would eventually bring apartmentalization, there are few apartment buildings in the neighborhood despite the relative old age of the neighborhood. The unwillingnes of most shop owners in trading land to the constructors for apartment construction can easily be traced from their narratives. The shop owner who owns one of the five apartment buildings in the region described the reason for the delayed apartmentalization process in the region as follows:

> The apartment construction remains very slow here. The most important reason here is that shop owners are residing on either side of the main road. Look at the apartment on the other side. They, themselves, built their apartment just like us. If they gave their land to the constructor, they got only one fifth or one sixth of the apartment. Since everybody is a trader here, we all know about financial issues a little bit. We are not satisfied with what the constructor offers.[398]

The unprofitability of having only two or three apartment flats in return for their lands is related to the rent earnings they are already taking from their real estates in the market. When apartments are constructed, the basement is designed to be shop spaces and the upper flats are used as family residents in general. Traders who own three or more small *gecekondu* shops on their lands may prefer conditions to remain unchanged since the rent revenues are greater than any returns they would get from the apartment construction. Traditional *gecekondu* shops are so small that three or four shops can be accomodated on the same plot owned by a shop owner, which constitutes a very satisfactory rent income. This finding of the research is supported by cases in many Third world cities, as stated by Davis[399], legalized squatting in Third World cities triggers slumlordism as a fact where the rent revenues from slum areas may be profitable when compared to investing in parts of the city.

---

398 Extract from field research note no. 10, 2 March 2003.

399 See Davis (2006:83).

In such an old *gecekondu* neighborhood that is close to important highways and where there is such a big market place, the apartmentalization should have already occured. However, the presence of shop owners as the main property holders of critical locations for construction, meant they had the potential to show a resistance to strategy's project of "spontaneously modernizing" these regions by means of apartmentalization.

Shop owners also describe another obstacle to apartmentalization that operated against their interests. The apartment construction necessitates relocation on part of shop owners at least for a certain period during the construction process. The shop owners, in general, describe their unwillingness to move out even for a short time because of concern about losing customers. In addition, if a plot of land is owned by more than one person, which was the case for some shop owners in the market, in order for a construction permit to be granted, all shareholders of the plot had to agree to the construction. According to the narratives of the migrants, in some cases, this constituted a barrier to apartmentalization. In general, the reasons behind the delayed apartmentalization reflect the increasing power of shop owners in determining the socio-spatial development in the region in relation to the legalization of the research setting in the late 1980s.

Some shop owners and other residents of Boğaziçi relate the belated application of the construction plan of the main road in the market place to the power of some shop owners in exploiting favoritism in local politics. The construction plan of the main road has been changed several times and it was not possible to fully apply it until the late 1990s. On my last visit to the region, the application of the plan was still not completed due to the resistance of certain shop owners, as mentioned by other shop owners. When you enter the market from Samsun Highway, the shops on the left have mostly been damaged and affected by the planning acts of Municipality. The *gecekondu* shops and houses in the market were built in a haphazard fashion and they do not follow a straight line. When the application of the plan is completed, parts of the shops and houses that exceed the line defining the route will be demolished. Some parts of the shops or the houses of traders were already severely altered in accordance with the conditions of the plan. Shop owners support a route either more to left or right, according to their shop location and therefore they evaluated the municipal acts of altering the road in accordance with the plan mostly referring to political favoritism.

There are two ways to explain this issue. First type of explanations defines the the application of the plan in its latest form leaning more on left as a result of technical necessity rather than as an act with political connotations. Second category of explanations, on the other hand, defines the decision and application of the municipality mainly as a political act resulting from the imposition of shop owners who have used their connections in the municipality. In most cases, the respondents signify the political connections that are against their own political affiliations, as will be described.

First type of explanations relate the construction of the main road leaning more on left rather than right edge due to technical necessities, for two reasons. First, the presence of public buildings like primary school and mosque on the right side is signified as the most important technical reason behind the latest construction plan of the road. Secondly, some technical explanations give references to the fact that the right side of the road is topographically suitable for construction when compared to the left side. Despite the fact that some shop owners suffered during the widening of the road themselves, they tend to define the process in a purely technical manner. These are the influential shop owners who define their relations with municipal planners and officials in a more active way when compared to others. The owner of one of the altered buildings, a member of "family X" explains the partial demolition of the house in accordance of the plan as a technical necessity and relates this to the topographically suitable nature of their side. The internal cross sections of their house became visible and the house seemed to become totally useless as such. The part of the house that was demolished can easily be perceived. In spite of this fact, the owner of the house mentioned his satisfaction regarding the proper implementation of the plan and the fair compensation received for their loss during the metropolitan mayorship of Melih Gökçek. The physical conditions restrict the possibilities for extension to the right, according to this house owner, and if the political connections played such an important role as people assume, they would have the power to exploit this fully, but, he added, this was not case.

In addition to dealing with the form of the latest construction plan many shop owners complain about the delayed construction of the road due to the favoritist policies of the previous municipalities. Many of the conservative *Sünni* shop owners who had no sympathy for favoritism operating to the advantage of social democrats or/and *Alevis* in the region blame social democrat municipalities for the delay. Some shop owners also point out that the construction plan had taken its final form during the social democratic metropolitan mayor of Murat Karayalçın. Therefore these respondents attributed the responsibility of their sufferings to the social democrat municipality rather than the recently elected municipality which has been applying the plan.

These different interpretations may at times reflect the actual facts, but most stories are filled with contradictory rumors and hearsay, revealing the traces of respondents' biases. A conservative *Sünni* shop owner blamed his sufferings on changes to the implementation of the construction plan carried out by the social democrat municipality. The recently constructed road passes through the land on which they owned a shop and house, which municipal officers demolished. He seemed angry at the incident, which happened in 1996. In part of our conversation, he lowered his voice and said that the road was not planned to pass across their land in the actual plan. However, in his opinion, later *Alevis* put pressure during Karayalçın's period and the plan was changed. According to him, in the later version of the plan, a slight arc was added to the road, causing it pass directly in front of their shop. He accused the owners of the bakery close to Samsun highway for that. The bakery was located just on the entry of the neighborhood from Samsun Highway. If the original plan had been applied the bakery should have been demolished. He claimed the bakers used their political influence to delay the implementation of the plan.[400] The owner of this bakery is similarly accused of the same thing by many shop owners. A *Sünni* shop owner from Yozgat explains the same incident as follows:

> We had a famous bakery here once just on the entrance of the market [He mentioned the fact that the bakery was located just on the place where the extension of the main road that once was river would be initiated]. They were the ones who had showed the main resistance to the development [with certain anger]. They were the reasons why no progress took place here for years.[401]

The owner of the bakery is from Karadeniz, though not *Alevi* in origin; he is said to have acquaintances in municipality when social democrats were in power. Therefore, *Sünni* explanations of the issue reflect the *Alevi* political connections with the municipality, and the favoritism typical of social democratic municipal tradition. An *Alevi* shop owner, on the other hand, interprets the same event differently, rejecting the accusations of social democrat nepotistic policies, and linking the belated application of the plan to the legal and financial barriers obstructing the municipal officers in dealing with the owner of the bakery as follows:

> At the entrance of the market [from Samsun Highway] there was a bakery just on the corner. This road could not be built properly because of this bakery until now. The bakery shop should be demolished according to the plan. The owners acted in a clever manner and insured their shops for very much above their real value. For years, therefore, the municipality could not afford to pay the insurance money in order to demolish the shop.[402]

---

400 Extract from field research note no. 2, 25 Sep. 2002.

401 Extract from field research note no. 34, 3 July 2003.

402 Extract from the field research note no. 19, 26 Apr. 2003.

This same shop owner, however, shared a similar criticism in relation to the improper attitude of the owner of the bakery causing the delay of plan's application. The construction of the main road constitutes a solution also to the serious problem of flooding. As a result of the delay of the road extension, the most recent flooding incident in the region had done a great amount of damage to the market, according to the shop owner. In this recent flooding, as a result of heavy rain, people in cars and buses were swept from the market to the Samsun Highway for about 200 meters. Shop owners were lucky to escape with their lives according to the stories.

There are cases where shop owners, who are densely represented in the neighborhood such as the ones from Kırşehir, are accused for delaying the progress in the neighborhood. Since they are the majority of shop owners and have powerful position as a group, they were said to take part in nepotistic politics, and therefore to be affected least from the drawbacks of the plan. A shop owner from Yozgat refers to this issue with the discomfort that "outsider" shop owners have felt for *Kırşehirlis* for years:

> Our place was two meters forward but we took it back [because of the construction plan]. The shop became smaller, very much smaller than in the past. We are not pleased with the Municipality. They did nothing, like to draw a proper construction plan [referring to the spontaneity of the plans]. The municipality is not helping us; they don't even have a concern like that. I have been here since 1967 and the progress should be more than this. We cannot construct our apartments, for example, we were eight people but my uncle died and we are 16 [share holders] now. There are so many problems related to infrastructure. Look for example; they could not even complete the construction of this road since 1983. There will be elections in ten months. As an example, the ones on the opposite side have acquaintances in the municipal council. They had already got all their compensation but they did not reduce the size of their shops. They are all from Kırşehir. Therefore there is no reason to expect any development here.[403]

The increasing power of shop owners is expected to define the implicit competition going on between *Alevis* and *Sünnis* after the 1980s through the production and use of space. It is interesting to see that, one of the primary areas of competition between them, namely the consumption sphere is dominated by the economic concerns rather than concerns related

403 Extract from field research note no. 34, 3 July 2003.

to sectarian or *hemşehri* identities. This will constitute the subject matter of next section.

6.6 Consumption Determinants: Prices or *Alevi/Sünni* Identities?

The increasing effect of neoliberal policies on the neighborhood space coupled with repression of active politics by state policies brought about the predominance of economic relations in employing and producing the space. The radically political space of the neighborhood in the late 1970s transformed into an economic space. The shops and commodities in the market proliferated qualitatively and quantitatively in the 1990s. In such a context, the once highly political categories of being *Alevi* and *Sünni* and other ethnic differences seem to have very little effect or seem to work as criteria for a limited group in defining the consumption choices. At least, both on traders' and customers' side, the tensions implicit were not made explicit in the expressions of the inhabitants and their portrayal of consumption habits. There were few incidents I witnessed during my stay at different shops making the tension explicit. The following anecdote by a *MHP* sympathizer member of the family X explicates such an implicit tension between trader and customer as a result of ethnic distinction:

> [After the formal interview]
> You should treat every kind of people well as a trader. For example, that man, a second ago was standing just beside me. You do not say to his face that he is a Kurd, *Tuncelili*, or anything else but you are giving him service. He has seen that we have been talking here, planted himself just beside me and sung a folk song [He tells these things with certain anger and strong prejudice.][404]

Interestingly, there are a number of examples as narrated by the shop owners and some inhabitants that material transaction between the people from the same sect or hometown may lead to an unfavorable positions for both sides. An *Alevi* shop owner defined his uneasiness with his *Alevi* customers in the following manner:

> -Are your customers mostly *Sünni* or *Alevi*?
> -To tell you the truth, the opposing group [he means that his customers are mostly *Sünnis*] comes more often. There is envy

404 Extract from field research note 16, 8 Apr. 2003.

> among our people. Our condition [he refers to his prosperity vis-à-vis other *Alevis*] is, so to say, good. They become jealous of our conditions. However, if something bad happens to them, who will run to help them? If something political happens, again we mobilize all our opportunities. The others [*Sünnis*] also know who we are but they like our honesty and kindness. And they have also had enough of their own people's hypocrisies; therefore, the majority of my customers are *Sünni*. More than that, our own people use the condition of being *Alevi*. They ask for money, you give; they do not pay you back. These things often happen.[405]

As well, from the perspective of the customers, the social proximity, in terms of being *hemşehri* or having common sectarian identity with the trader, may not be a favorable condition when a difficulty in bartering emerges. An old couple narrated the difficulty in the following manner:

> We prefer not to buy anything from an acquaintance. Then they cheat you as a friend. It is far worse than shopping from a stranger. Other than that, you cannot barter with a friend. Shopping from a stranger is better. You can put the price as you like.[406]

From the traders' perspective, sectarian identities hardly seem to constitute a serious concern. Having trustworthy and reliable customers is the most important criterion for shop owners when selling goods on credit. In this sense, hometown commonalities play a minor role in the shop owners' decision to sell goods on credit to certain customers.

From the customer's perspective, on the other hand, economic concerns, spatial proximity of the shop and the possibility of shopping on credit dependent on having known each other for a long time seem to have a strong effect on consumption choices. Particularly in relation to food shopping, almost all of the respondents mentioned their preference for big supermarkets that opened in the market beginning in the early 2000s. The supermarkets were defined as more suitable for the respondents, mainly because they could use credit cards there. The super markets, in competition with each other, sell some goods, particularly food, below real prices, taking a risk of loss. Therefore, mainly for food shopping they are seen as the best alternative for the migrants. The *bakkals* and old grocery stores have suffered most from the opening up of these super markets. They defined their incompatibility with the markets in the following manner:

405 Extract from field research note no. 19, 26 Apr. 2003.

406 Extract from field research note no. 36, 20 July 2003.

They started business here two, three years ago. They, of course, put an impediment before our job. I went to a wholesale food market to buy goods for the shop. We chatted with the directors of Nazar and Gima [The supermarkets that have branches in different locations of Ankara city]. They said that they bought one kilogram of tomato for 600,000 [approximate equivalent to 50 cents] Turkish Liras and sold it in the market for 500,000 Turkish Liras. They said: "We sell tomatoes and cucumbers at a loss and regain this loss in other goods in the market". We cannot compete with them. Is it possible? I am asking you.[407]

For their other needs like clothing and house appliances, the migrants go to shops with which they have developed some kind of trust and intimacy over the years. They mainly buy things on credit from these shops. Traders mentioned their function as an economic buffer mechanism by delaying the debts of customers for a number of months. Because of this, shopping and selling on credit is still a norm in the paradoxically modernized space of the market, it is a tactic on the part of shop owners to compete with the big supermarkets and shops in other parts of the city. Poor people in general have difficulties in paying their debts to shop owners. Particularly when the shopping for medicine is the case, the condition of poor people become more obvious. The owner of the most visited pharmacy gives remarkable examples about the issue:

-Do you sell medicine on credit?
-[The pharmacist showed me a box in which there were pieces of jewellery and identity cards that had been left as guarantee by the customers in return for their debts] Here there is an identity card and a driving license left a year ago. He left the driving license is a guarantee for a 3.5 million Turkish Liras debt. If they want to get a new driving license they will have to pay much more. They, in some cases, have left their earings. I took them to the jewelry shop next door. They said that they are copper.[408] [She laughs.]

Moreover, the inhabitants, for the sake of not severing their close relations with *bakkals* they have known for a lengthy period may, at times, shop from them in order not to upset them. The respondents, in some cases, related their emotional discomfort vis-à-vis these *bakkals* because of their choice of

407 Extract from field research note no. 16, 8 Apr. 2003.
408 Extract from field research note no. 18, 25 Apr. 2003.

supermarkets for food shopping. They do not want to offend the *bakkals*, not only for the sake of their close personal relations with them but also for the financial comfort they provide in times of financial difficulty. The continuity of the old and trustworthy relations with *bakkals* constitutes a guarantee for the inhabitants in various ways. This trust relation between the customers and *bakkals* can be well exemplified by the report of a *bakkal*:

> People generally go to supermarkets for their collective shopping needs. The ones who do not have money and have minor needs come to us. The ones who shop from me are all acquaintances. Yesterday a woman who had shopped from Özsoy [a supermarket in Boğaziçi] came to me with her packets in hand. She said: "Haydar I did something stupid. I shopped from the market. Please calculate what I have bought so that I can see if I was cheated". I said that I could not do something like that.[409]

In spite of the fact that the market place has been "modernized" with the integration of a neoliberal economic space in the neighborhood, the traditional way of economic transactions continues to play a vital role for the inhabitants, particularly for the needy. Even though these newly opened shops appear to affect the business of traditional shops in a negative way, since old shops perform an important function as a buffer mechanism against times of crisis, they continue to exist in the market.

6.7 Revitalization of *Alevi* and *Sünni* Identities: The Politics of Religious Spaces

As mentioned above, the military regime's efforts to propose a *Sünni* interpretation of Islam as a remedy against leftist ideology and Marxism not only strengthened the popular appeal of Islam in low-income settlements in big cities and in provinces where *Alevis* are densely integrated with the *Sünni* population, but also led to the revitalization of ethnic and sectarian identities that had always been in minority position. Islam became one of the main components of the ruling ideology operating as a bulwark against the leftist movements of the 1970s. The approach of the state towards *Alevis* in particular has always reflected a hesitant position. From the establishment of the Republic onwards, on the one hand, the secular basis of the state has always been marked and has constituted the reason for most *Alevis'* loyalty to Atatürk and his party *CHP*, but on the other hand, *Sünni* Islam

409 Extract from the field research note no. 7, 16 Oct. 2002.

has been placed in a privileged position vis-à-vis other interpretations of Islam. As pointed out by Göner[410] the dominant public perception in Turkey paradoxically excludes Islamist tenets and, at the same time, groups who define themselves outside of *Sünni* Islamism. The Presidency of Religious Affairs, a state institution, considers only *Sünni* Islamic religious needs and practices while ignores all other interpretations of Islam including *Alevism*, despite their duty in principle to provide religious services to all Muslim citizens without discrimination.[411] The official approach to *Alevism* is obvious in the book "Alevilik Bektaşilik Debates" prepared by the Presidency of Religious Affairs. According to this source, certain premises summarizing the state's official ideology about *Alevism* can be defined as follows: *Alevism* cannot be thought of as an independent religion or extension of other systems or ethnicities; *Alevism* cannot be thought as an Anatolian interpretation of Islam and *Alevism* cannot be identified with communism.[412]

As a consequence of the hesitant approach of the state towards *Alevism*, the serious conflicts and fights between *Alevis* and *Sünnis* around leftist and rightist ideology in the late 1970s and the rise of postmodern discourses of multiculturalism and identity politics after the 1980s, there has appeared a "re-politicization of Alevilik in the 1990s".[413] As argued by the authors, this re-politicization is qualitatively different from the pre-1980 politicization of *Alevis* as part of the Socialist movement. Recently, *Alevi* communities are emphasizing their identities more with reference to religious and cultural qualities. In this sense, the previously radically political identity claims that were identified with the class issues and economic deprivation of *Alevi* groups has been replaced with religious claims and an understanding of *Alevism* as an egalitarian world view and way of life. "In an interview at the Karacaahmet Foundation, Vahap Güngör argued that "when the axis of politics was left-right we (*Alevis*) chose to be on the left; but now the axis of politics is Shari'a versus Laicism and we choose to be *Alevis*".[414] This fact also has an important connection with the rising *Sünni* Islamic fundamentalism and long-lasting and assimilating discourses of *Sünni* Islam.

In parallel to these developments and because of the hesitant strategy of the state regarding identity politics, *Alevi* and *Sünni* identities have to a large extent lost their explicit political connotations in the narratives of the

---

410 Göner (2005:119).

411 Ibid., p.114.

412 Quoted from Yılmaz (2005:177).

413 Erman, Tahire and Göker, Emrah (2000) "Alevi Politics in Contemporary Turkey," *Middle Eastern Studies*, 36(4), p. 100.

414 Göner (2005:125).

respondents, when they refer to the late 1980s. They only regard sectarian identities as political in the context of favoritism, and not out of the context of left and right. In general, the religious content of *Alevism* and *Sünnism* has dominated the respondent's narratives in relation to recent incidents of social spatialization.

The way the public space is used and produced by these two groups has, to a large extent, also been transformed. As mentioned in previous sections, the boundaries and competition between *Alevis* and *Sünnis* in a public space that was once radically political, has become more embedded in the production and employment of space. Being *Alevi* and *Sünni* seems to explicitly dominate the narratives of both sides when talking about the spaces and life styles of the "other". *Alevi* and *Sünni* identities, in terms of their recent connotations for the migrants, define the "other's" religion rather than their political identities of left and right. In this sense, despite the suppressive policies of state, inhabitants do not neglect an awareness of their group identity and preserve a certain degree of tension and uneasiness towards the "other" group. Akyüz[415], who conducted research in the Boğaziçi neighborhood on religious life and modernization, argues that there exists a sharp distinction, and awareness about *Alevi* and *Sünni* identities among migrants. The percentage of the people who define religion as the main factor behind their choice of friends and social interaction is 53.6%, which is considered quite high.[416]

Depending on the context of the relationship between the *Alevi* and *Sünni* migrants, the definitions of the "other" may show variations. In cases where *Alevi* and *Sünni* families have had a long history of neighborhood relations with each other, the uneasiness in the narratives of the respondents is, at times, replaced by praise of the good attitudes of their neighbors, despite the fact that they are *Alevis* or *Sünnis*. Particularly for *Sünnis* who carry relatively more prejudices and superficial codes of *Alevi* people, the appreciation of *Alevis*, in most cases, depends on having known each other for so long as neighbors, as is the case in the anecdote below:

> Neighbors cause no harm to each other. This is a rule. In this region people are living densely, close to each other [he means the heterogeneity in terms of sectarian identity]. These people [he means *Alevis*] are open-minded when compared to our

---

415 Akyüz, Niyazi (1994), "Ankara'nın Boğaziçi Semtinde Dini Hayat ve Kentlileşme Üzerine bir Araştırma", Unpublished Ph.D Thesis, Ankara Üniversitesi Sosyal Bilimler Enstitüsü: Felsefe ve Din Bilimleri Anabilim Dalı, p.157.

416 See Akyüz (1994:158).

> people. My relatives living there [he means among *Alevis*] have better neighborhood relations. There is no jealousy among them. They do not have so many children. They are certainly more tolerant.[417]

However, even in these cases, the recognition and prejudices of the distinction between the *Alevi* and *Sünni* lifestyles and the undeserved qualities of each lifestyle dominates the expressions of respondents. In cases where the trust between *Alevi* and *Sünni* neighbors was seriously damaged in the late 1970, or through the informing activities in the early 1980s, the relationships between neighbors was at times, totally broken. In the narratives of both *Sünnis* and *Alevis,* one can sense the act of "othering" implicit in their comments when talking about the people from the opposite sect and their undeserved or negative qualities. Particularly *Sünni* respondents reflect their prejudices combined with a neglectful and ignorant attitude about *Alevism*. The superficial codes about the *Alevi* faith in the expressions of *Sünni* migrants are frequently present and obvious. During the research, the general approach of *Sünni* migrants towards *Alevism* was often voiced with a noticeably superficial overtone and knowledge about the *Alevi* belief system and with certain degree of ignorance. This fact has been supported by the findings of many studies.[418] As Göner points out, referring to the reports of an *Alevi* interviewee, *Sünni* neighbors refuse friendly offers from her family on the grounds that "the meat *Alevis* offer is not edible".[419] These prejudices in everyday life at societal level, according to Göner, operate as crucial informal mechanisms by which the access of *Alevis* to political and economic resources is blocked. Some *Sünnis* in Boğaziçi neighborhood identify the non-practicing of *namaz* and *oruç* (fasting) during Ramadan among *Alevis* with atheism as pointed out by the findings of Akyüz.[420] It is necessary to note that these prejudices are, most of the time, present but not explicit in the narratives of the respondents.

In the narratives of the respondents, these negative value judgments exist in a hybrid form with their pleasant, personal experiences with *Alevi* friends, reflecting the inconsistent state of mind. Under all conditions, the

417 Extract from field research note no. 29, 20 June 2003.

418 See Çakır, Ruşen (1996), "Political Alevism versus Political Sünnism: Convergences and Divergences" in Olsson, T., Özdalga, E. and Raudvere, C. (eds.), *Alevi Identity*, İstanbul: Swedish Research Institute in Istanbul Transactions, p.64;.Göner (2005); and Akyüz (1994).

419 Göner (2005:114).

420 Akyüz (1994:149).

narratives contain an awareness of difference between *Alevi* and *Sünni* life styles. The *Alevis* who accept and observe some of the religious obligations of the *Sünni* belief system, like hac (pilgrimage) and *namaz*, are appreciated in such often-encountered expressions. The narrative of a *Sünni* woman from Kırşehir well explicates this inconsistent state of mind. She wanted to emphasize her neutral position vis-à-vis *Alevis* but gave hints of her prejudices and real approach in her narrative in the following manner:

> Around here there are not so many *Alevis*. *Kırşehirlis* are living here... People from Sivas are generally *Alevi* in origin. The drummers no longer go there [the region where people from Sivas are densely populated] during Ramadan.[421] They started beating the drummers there. They do not fast, you know. You are not *Alevi*, are you? [She asked the question with a certain concern about behaving shamefully. I said "no".] Neighborhood relations have been very good, though. I like the *gecekondu* life because of the neighborliness. Nobody knows each other in apartment life. *Alevis* of Yozgat and Sivas reside here. Among them, there are people who have been to *hac* and a number of them also go to the mosque and pray. The number of these may increase but their people react so badly. I have a relationship with a number of them. Actually we are close to all of them. We do not discriminate. In my opinion, these men who visit the mosque become Muslims.[422]

This anecdote is typical of the conservative *Sünni* approach that I encountered a number of times during the research. The story of the *Alevi* beating of the drummers in Ramadan and their non-practice of *Sünni* Islamic obligations was narrated by a *Sünni* woman as proof to their non-believer position in an attitude that showed total ignorance of *Alevi* rituals. The commonly held belief about *Alevis* as being non-Muslim is quite in compliance with the hegemonic discourse of a state that heightens *Sünni* Islam as the only legitimate interpretation of Islam. This is the reason why, in the 1990s,

---

421 Ramadan is the holy month in the *Sünni* Islamic belief system. Muslims fast in the daytime during Ramadan. In order to feel satiated, they wake at night and eat something before daylight according to the rules of the religion. Drummers, traditionally walk along the streets at nights to wake people up during Ramadan. This tradition is still practiced in cities in Turkey. *Alevis* also fast, but at a different time and form than *Sünnis*.

422 Extract from field research note no. 39, 15 Aug. 2003.

within the context of the re-politicization of *Alevism*, the *Cem* foundation, a broadly based *Alevi* organization, aimed at an acceptance of *Alevi* culture and beliefs. They want the state to allow the same kinds of activities in the month of Muharrem,- regarded as the holy month by *Alevis*-, as it does in Ramazan.[423]

Alevis also accuse *Sünnis* of malpractice of *Sünni* Islam and for being hypocritical when practicing their rituals. Akyüz[424] mentions that *Alevis* mainly criticize the formalistic nature of *Sünni* prayers and points to the inefficiency of praying five times a day, wearing a headscarf, or fasting if your heart and mind is full of immoral and inhumane ideas and emotions. An *Alevi* shop owner, while talking about the *Cemevi* in the region and *Cem* meetings, gives references of the hypocritical attitude of *Sünnis* in the following manner:

> In the past, the rules of *Cem* were harsher, now they are not like that. For example, they [*Sünnis*] fast; but at night they sleep with their wives. Our fast is not like that; you cannot even drink water. During the fast, rather than sleeping with your wife you cannot even share the same bed. For example, this Halitli village association they are from *MHP*. They fast during Ramadan. Then they gather at nights in the association and gamble until four in the morning. I don't understand their way of fasting. Another thing, everybody can take part in *Cem* meetings according to our rituals. A third person [person who is not *Alevi*] or women and men together... Can you discriminate against people in front of Allah? But they discriminate against people as man and woman.[425]

According to my observations in Boğaziçi, the production and use of religious space constitutes the most important issue, where *Alevis* and *Sünnis* continue to signify their differences from each other within the context of revitalization of identities vis-à-vis each other. In other words, the production of religious space for both sides has become the main means of preserving their identities. The production of religious space is mainly considered the responsibility of the community rather than the state; in this sense, the informal mechanisms through which these spaces are produced mainly depends on the political and social networks that these groups have,

423 Göner (2005:130).

424 Akyüz (1994:149).

425 Extract from field research note no. 19, 26 Apr. 2003.

vis-à-vis each other. As pointed out by Erder[426] in her study on a low-income settlement, it is considered by migrants, in general, that some services are the responsibility of the state while others can be a product of cooperative work or should be totally solved by the community, depending on the accounts of *muhtars*. In the *muhtars'* understanding, the laws of the Turkish Republic give the responsibility of building schools to the state, while the building of the mosques is the responsibility of the community. In this respect, according to Erder, many issues, from the construction of mosques to sustaining housing, are considered as community responsibilities while service provisions related to transportation, the sewage system, electricity and schooling are considered as the responsibilities of the state.[427] This is the case also in Boğaziçi where the mosques and *Cemevi* in the region had been built as a result of community effort.

State policies encourage *Sünnis* to build mosques, while construction of *Cemevi* as a space of prayer was still considered as illegal despite the pressure of EU in relation to *Alevis'* rights to freely practice their rituals. The military regime had followed policy of suppression and assimilation of *Alevis* by means of opening mosques in *Alevi* villages and calling upon all *Alevi* to attend worship in the mosque in the 1980s. The *Sünni* Islamists' strict reduction of Islam to a single interpretation explains their support for such assimilationist policies of the military regime.[428] From the mid-1990's onwards, the domination of religiously oriented parties in local politics had further facilitated the conditions of religious practices and rituals for *Sünni* residents in low-income settlements. This dramatic change in the political preferences of the *gecekondu* settlers also affected the approach towards these settlements by the "urbanite" sections of society, who have appropriated to a large extent modernist, republicanist and, therefore, strongly secularist tenets as mentioned in chapter two. From this tension, there emerged a divide between secular versus Islamic groups replacing the leftist and rightist divide of the 1970s. The urbanites' concern for secularism led to a hardening of the exclusionary discourse. Meanwhile these developments have instigated the spread of political Islam in everyday life in *gecekondu* settlements. Local religious communities and gatherings increased in number, particularly in the 1990s. The illegal tariqat[429] (tarikat) organizations and closed religious

---

426 Erder (1997:123).

427 Ibid., p.123.

428 Çakır (1996:64).

429 *Tarikat* is a word of Arabic origin, which is defined as 'a method of moral psychology for the guidance of individuals directing their lives toward a knowledge of God' in The Encyclopedia of Religion, 1987.

community formations have attracted considerable attention from the public and this has also contributed to urbanites' anxiety.[430]

As concluded by Akyüz[431] in his findings on Boğaziçi neighborhood, the percentage of migrants who explicitly mention membership of a tariqat as beneficial for proper practicing of Islam is 37.1%. This is a rather high percentage, taking into consideration that tariqat membership is forbidden by law. The intense surveillance of security forces on illegal *zikir*[432] gatherings during the period of social democrat dominance in local politics was loosened in the late 1990s as learned from the informal conversations with the respondents. Other than that, it is important to note that, I personally encountered many religious gatherings[433] among women and heard the voice of *zikirs* arising from the houses so frequently during walks in the neighborhood. *Sünni* women discussed their sufferings related to both home meetings and meetings at the masjid during the governing of social democrat mayors. In recent years, the religious meetings, particularly at certain houses that had been previously raided have been watched over by the gendarme forces. The intense supervision of such gatherings by state security forces led to a perception of illegality among the inhabitants of some gatherings. Therefore, some *Sünni* women prefer not to go to gatherings at certain houses, rather they themselves organize smaller-scale religious meetings in their own houses. When compared to *Alevis* though, it is

---

430 See Demirtaş and Şen (2007:97).

431 Akyüz (1994:150)

432 The common essence of all tariqat organizations is *zikir*. Tariqats teach Muslims mostly how to practice *zikir* in order to be close to Allah by reciting his names in a right way. See Akyüz (1994:147).

433 They mainly prefer their neighbors from their hometowns in order to make these gatherings close to outside. The participatory observations made in two religious gatherings show that these gatherings perform two important functions for the women. First of all, they talk about religious issues and try to find answers to their questions about daily issues via the religious teachings. They also feel released from their everyday problems with the comfort of worship. Secondly, these religious gatherings take the place of *gün*, which is the traditional way of regularly coming together for women in Turkey. During the whole ceremony, all the windows were closed despite the fact that the weather was so hot. In some part of the conversations, they have mentioned their concerns about unexpected attacks. They said that similar meetings in the neighborhood happened everyday. The continuous supervision of women gatherings by state security forces has led to a reservation on part of women. They give extreme importance to the privacy of these meetings and it is very difficult to take part in these gatherings as a researcher.

possible to argue that *Sünnis* express their demands and beliefs more easily and comfortably. In that respect, as observed in the research setting, *Sünnis* seem to feel quite comfortable vis-à-vis *Alevis* and local political actors in making their religious claims to public space explicit. As also proved by the findings of Erder's[434] study, *Sünni* communities meet no difficulty in sustaining local aid for the construction of mosques and Quran courses in the neighborhood, unlike the *Alevis*, who found it impossible to obtain for their places of worship.

Despite the fact that there is no reference to an open competition over space in religious terms, there is a certain dislike showing itself in the narratives as an intimidating lack of awareness or neglect about each other's religious spaces. The ignorance and neglect on part of *Sünnis* about *Alevi* society and belief system agitate *Alevis* most as understood from the narratives of the respondents.[435] This neglect on part of *Sünnis* has gone together with the continuous dominance of *Sünni* Islam and its obligations to *Alevi* population. While *Sünni* respondents seem to be indifferent to the *Alevi* belief system and *Cemevi* as the space of *Alevism*, they expect *Alevis* to play roles in the production and recognition of *Sünni* Islamic space. The construction of mosques in the region clearly reflects these expectations of *Sünnis* from the *Alevi* residents of the region. As mentioned above, mosques and *Cemevis* are built with the donations of the community in *gecekondu* neighborhoods. The Mevlana mosque on the upper parts of Boğaziçi neighborhood just on the boundary between Boğaziçi and Akşemsettin has a construction history that can constitute a good example for this process. This area is heavily populated by *Alevis*. Despite the fact that *Sünni* inhabitants are aware that *Alevis* do not pray in the mosques, both *Sünni* and *Alevi* respondents made reference to the construction committee's pressure on *Alevis* to take active part in and donate the construction activity. Listening to the history of the construction of the mosque from the head of the construction committee and from *Alevi* residents living close by reflects both side of the story.

---

434 Erder (1997:130).

435 "From time to time the former Welfare Party (*WP*) made sympathetic calls to *Alevi* communities and invited their "religious brothers" to the party ranks. However, especially after the Sivas event and its aftermath which witnessed the pro-event reactions of *Sünni* Islamist media and *Sünni* politicians, as well as the anti-*Alevi* and / or assimilationist declarations of the high-ranking *WP* members, a great majority of *Alevis* today, regardless of their political dispositions have become highly suspicious of the pro-*Sünni* politics." Erman and Göker (2000:109). In 1993, *Sünni* Islamist residents of Sivas committed a massacre to participants in an *Alevi* cultural festival by setting their hotel on fire.

Mevlana mosque has no minaret and this is evaluated as a serious problem both by *Sünni* and *Alevi* people residing close to the mosque for different reasons. The head of the mosque construction committee related the unfinished view of the mosque without minaret to the dense settlement of *Alevi* population in the neighborhood. He mentioned the difficulties of collecting donations, which had mainly taken place in the coffeehouses as the most important public places. Owners of the coffeehouses are people from Yozgat and a mainly Yozgat *hemşehri* community seems to use these two coffeehouses. In this respect, the head of the construction committee indicated coffeehouses as the most suitable places to collect donations for the mosque. People give their alms during Ramadan, a requirement in *Sünni* Islamic practice in the form of donation to mosque construction. However, despite the donations of community, according to the words of the head of construction committee, there was a need to travel to big mosques in the city to collect further donations. Still these attempts were not sufficient according to him because of the dense presence of *Alevi* people in the neighborhood. He relates the difficulty of collecting donations in the neighborhood mainly to this fact in the following manner:

> We have furnished the mosque with green carpet lately. We bought catalytic stove. It gets cold from time to time. We are going to construct a natural gas heating system. For the time being we put a boiler and heat the mosque with oil. In the past, we had gone everywhere with our receipts in our hand. We went to Ulus, to the big mosques there. It is very difficult to collect donations here; there are so many *Alevis* in this neighborhood. The reason for the mosque looking as it is [without a minaret] is mainly because of that.[436]

Sünni respondents in the region relate the matter to the *Alevis'* unwillingness to donate and their lack of Islamic faith. On the other hand, the scene of the mosque without minaret was explicitly linked by *Alevi* inhabitants to the possibility of the abuse of donations or inefficiency on the part of the mosque construction committee. They claimed that they have donated the mosque financially on a continuous basis but still the mosque has an unfinished appearance, which was a shame. The following anecdote seems to illustrate the ideas of *Alevis* well:

> -Have you ever donated to the mosque construction committee?
> -Of course, the ones in the committee and the hocas (hodjas)

436 Extract from field research note no. 53, 3 Sep. 2005.

> collect money. We had given money for several times. But look at the mosque. It has no minaret. Have you ever seen a mosque without minaret? Is it possible? To give an example, if I describe my address to someone as the house near the mosque, nobody can find the house, because there is no minaret. Since 1973 there has been no minaret. They collect money all the time. Can't one construct a minaret in 32 years?[437]

Besides that, *Alevis* mentioned their awareness of the fact that *Sünni* people perceive the main reason behind the lack of minaret as their dense presence in the region, which has made them quite outraged over the issue. They describe their feelings of being treated unjustly by the committee members despite the fact that they had donated to the mosque several times. On the other hand, *Sünnis'* answers to questions related to *Cemevi* reveal their ignorance of, lack of interest in, and unease over the issue. Regardless of the fact that I was well aware of the answers I would receive, I asked some *Sünni* inhabitants if they had helped by any means to the construction of *Cemevi* in the region, in order to gauge their reactions. They appeared to be quite shocked at the question, despite the fact that they perceived *Alevi* donations to the mosque as totally normal and expected.

Similar to the findings in Boğaziçi, in Erder's[438] study, the respondents who define themselves as *Sünni* accept *Alevis* mainly as a *hemşehri* community; however, they were quite reactional and opposed to the explicit practicing of *Alevi* rituals in public. In that sense, the opening of *Cemevi* in their neighborhood seems to constitute a big concern for them. *Alevi* groups, therefore, if organized around their community are met with cultural, political and administrative obstacles in most cases according to Erder. This finding also holds true in the case of Mevlana mosque.

The construction of *Cemevi,* in that sense, was narrated to be a product only of *Alevi* efforts rather than a community attempt in the neighborhood like mosque construction. The construction of only *Cemevi* in the region was initiated in 1995. Despite the dense presence of *Alevi* population in the region, there was only one cemevi for a quite large area. The cem meetings were made by turn according to the hometowns of migrants. People from Yozgat, Tunceli, Sivas and Çorum all took part in their own ceremonies in winter, each with their own *dede.*[439] People from Çorum and Yozgat were said to

437 Extract from field research note no. 55, 4 Oct. 2005.

438 Erder (1997:136).

439 *Dede* (Holy men) is religious leader who belongs to a hereditary priestly caste and who directs cem ceremonies. Every village has its own *dede.*

come together more often, because their *dede* is living in the neighborhood unlike the other *dedes* who live back in the village. Other than the cem meetings in winter, *Cemevi* seems to be an empty space. As understood from the narratives of *Alevi* respondents, *Cem* meetings are rarely performed and *Alevi* religious practices seem to be more fragmented when compared to *Sünni* collective action in the production and employment of their religious space. The fragmented nature of *Alevi* society's demands and interests constitute the most important obstacle behind the difficulty of community action in neighborhoods like Boğaziçi.

The fragmented natue of *Alevi* society stems, to some extent, on the inherent qualities of *Alevi* rituals and belief system. The rituals of *Alevism* necessitate, in some cases, the inclusion of close community as a result of its need to preserve the privacy of the group. As pointed out by Okan[440] the reasons behind *Alevis'* not permitting *Sünnis*, or even *Alevis* from different villages to participate in their *Cem* meetings emerge from their belief that the problems of the community should be kept private. Furthermore, *Alevis*, depending on where they live, may hold different positions in regard to the interpretation of Islam. For example, the *Alevis* of the villages in Çubuk and Beypazarı or the towns of Ankara, perform the *namaz*, therefore accept a more *Sünni* version of Islam, while the Kurdish and Turkish *Alevis* of Sivas do not approve and distance themselves from these practices. This fragmentation can also be followed from the associational divide and claims of *Alevi* organizations in public space in the 1990s. The perception of two broadly based *Alevi* associations has greatly divergent positions with regard to the interpretation of Islam. The *Cem* foundation adopts a definition of *Alevism* as "Turkish Muslimness" while the Pir Sultan Abdal Foundation mainly regards *Alevism* as a secular and democratic way of life rather than as a religious identity. The European *Alevi* Union Federation and Hacı Bektaşi Veli Foundations seem to find themselves an intermediary position between these two understandings of *Alevism*.[441] *Alevi* respondents, during the conversations clarified their understanding of *Alevi* belief system and they tended to show such variation, and it was noticeable that former radical leftist militant *Alevi* embrace a form of *Alevism* similar to the standpoint of Pir Sultan Abdal or Hacı Bektaşi Veli Foundations. Among the more conservative *Alevi* groups, groups of Çorum and Yozgat, the view that *Alevism* is a Turkish form of Islam seems to be widespread. The fragmented nature of the search of *Alevi* identity after the 1980s has also to do with the deterioration of traditional *Alevi* culture and rituals such as *dedelik* with the urbanization of *Alevism*. The *dedes* lost their important role in the city environment particularly as a result of the increasing

440 Okan, Murat (2004), *Türkiye'de Alevilik* (*Alevism* in Turkey), Ankara: İmge, p.74.
441 See Okan (2004:125).

communication facilities. Furthermore, as a result of the spread of Marxist leftist ideology among *Alevis* and the class understanding of leftist ideology, *dedelik* began to be considered as a part of the exploiter-exploited relation.[442] The fragmented nature of *Alevi* society is clearly revealed by their inability to unite under the roof of one political party. The former leader of Turkish Union Party, (Türkiye Birlik Partisi-TBP) established in 1966 with the objective of representing *Alevi* society, Mustafa Timisi connected the lack overall *Alevi* support for the party to the distance of *Alevis* to the idea of a political party that unites all *Alevi* interests.[443]

The fragmented nature of *Alevi* society when encountered with the hegemonic domination of *Sünni* Islam and society, it becomes understandable why *Alevi* community in Boğaziçi cannot build and use *Cemevi* as easily as *Sünnis'* produce and employ their religious spaces. As mentioned above, there is also a legal barrier to the construction of *Cemevi* and this fact has been severely criticized by *Alevi* organizations in recent years. Even in cases where *Cemevi* projects have been presented as social and cultural foundations rather than places for prayer, there is a great possibility of recently appointed local officials denying permission to build on certain pretexts. This is the case also in Boğaziçi. For this reason, *Cemevi* has an unfinished appearance and according to the statements of the head of the construction committee who is also close relative of dede of Çorum, municipal officers placed serious legal impediments in front of them. They still struggle with police officials in order to prevent the building from being demolished. He reported the legal impediments in the following manner:

> We experienced great difficulties in taking construction authorization. We gave a coal cellar as our address. We cannot perform a legal opening. They [municipal officers] said that it was built on the land of *ASKİ* (Ankara Water and Sewer System Department). We would like to go to the municipality in order to put our claims but they always postpone our meeting. ASKİ is near the police station and besides them is the big water depot. This area is seen as the green area in the construction plan. During the municipal period when *CHP* was in power in Mamak, these areas were all legalized and opened to construction. But they want to put impediments before us [as in the form of setting forth the land of *Cemevi* as in the status of *ASKİ*'s land].[444]

442 See Yılmaz (2005:149).

443 See Schüler (2002:163).

444 Extract from field research note no. 50, 5 Aug. 2005.

In spite of the fact that the presence of *Cemevi* and its construction has a highly symbolic meaning and importance for *Alevi* people, the space seems to be an empty one in everyday life except for occasional meetings conducted for at certain periods in the year. This is in some way connected with the illegal status of the space at the moment. However, more than this, these places gained attention and extreme importance in times when the manifestations of identities was needed. In this sense though, *cemevi*, masjid or village associations may not be considered as lively, "full spaces", they stand as the spatio-symbolic indications of *Alevi* and *Sünni* or *hemşehri* identities. Even for those who are not fully committed to *Cemevi* rituals, the difficulties of building a *Cemevi* and the possibility of its demolition constitute an important concern. Therefore, it is possible to understand through a spatial perspective that *Alevi* and *Sünni* identities as mainly religious identities seem to protect their implicit importance after the 1980s. The increasing domination and imposition of *Sünni* Islam over everyday life have led to the revival of *Alevi* and *Sünni* identities. The religious recognition on both sides has affected the neighboring relations and particularly in the way women have congregated and formed relations with women from the other sect. Because of being neighbors and due to spatial proximity, *Alevi* and *Sünni* women may meet each other; however, they do not visit each other specifically. Most of the time, women's gatherings include the women from the same hometown or from the same religious sect. This is not a mere coincidence, since most of the time *Sünni* women's gatherings are for the purpose of religious sharing, as mentioned above.

6.8 Conclusion

The course of social spatialization in the neighborhood beginning in 1980, in general, has been affected by the "depoliticization policies" of the military regime and the succeeding civilian governments. The suppression of active politics by means of close surveillance of security forces in the locality, the general imposition of neoliberal political agenda and the strengthening of *Sünni* interpretation of Islam over *Alevi* can be considered as the most important components of this depoliticization strategy. The suppression of active politics by means of security forces' close surveillance and the emergence of an informer network among migrants have further alienated *Alevi* and *Sünni* migrant communities from each other.

The neoliberal economic policies, on the other hand, had broken down the community claims to a large extent and replaced them with the economic concerns of individuals. The policies have also started an immense legalization movement in *gecekondu* settlements. This legalization process has been operationalized via a reform of decentralization in local government system and the operation of private construction offices. Neoliberal policies seek the "spontaneous apartmentalization" in these settlements by making *gecekondu* lands subject to market mechanisms and to the interests of big construction firms as attractable peripheral lands.

However, the loopholes of the neoliberal legalization strategy that was open to abuse both by migrants and officials; the increasing local agency of migrants with the growing favoritism in local government; the complex nature of land ownership patterns and the increasing reciprocity and responsiveness between tactical and strategical realm, in general, led to the emergence of unforeseen consequences in spatial terms. Delayed apartmentalization in the designated research setting can be given as an example to these unintended consequences. In congruence with the growing reciprocity and responsiveness between strategical and tactical realms, there has emerged a transition from "space of politics" to a "politics of space" in the locality. In other words, the radically political nature of the public space in the late 1970s has been replaced by an implicit competition between *hemşehri* and sectarian groups that has been embedded in and defined by diverse claims over the social spatialization of the research setting by the 1980s.

The third component of the depoliticization strategy, that is, the domination of *Sünni* Islam over *Alevi,* has led to a general revitalization of *Alevi* and *Sünni* identities mainly in religious and cultural terms. This revitalization of identities has had a serious effect on the nature of social spatialization in the locality. The widespread practice of *Sünni* Islam in the research setting, particularly after the mid-1990s, has brought about a perception of being threatened on part of *Alevis*. In that respect, besides economic claims, religious claims to space have determined the course of social spatialization beginning by the mid-1990s.

Unlike the nature of spontaneity that was mainly a consequence of the minimum or non-intervention of strategical realm in the 1970s, the spontaneity of social spatialization in the last two decades has been a consequence of the growing local agency that is able to perform near-strategical acts and the increasing reciprocity between the strategical and tactical realms on the vertical level. This reciprocity, however, has functioned mainly to the advantage of these powerful groups having the capacity to perform near-strategical acts rather than the migrants, who ought to behave tactically within the context of strategical realm on the vertical level and near-strategical realm on the horizontal level.

CHAPTER 7 - Conclusion

The main theme of the book is social spatialization in the locality. Individual stories of migrants are the constitutive ingredients of this story. They illuminate certain critical moments of social spatialization that are considered crucial for this research. The main points of research findings will be elaborated in this section. After these general empirical conclusionary remarks, possible theoretical contributions of the study will be emphasized.

The story of social spatialization began in the 1940s, the times where the first encounters of forerunner migrants with the locality took place. Aside from the macro-political, social and economic reasons behind the rural-to-urban migration, the emergence of a *gecekondu* neighborhood in this particular locality had been closely related to the early planning strategies of Ankara city, with unforeseen consequences. The main reason behind these consequences results from the irreconcilable conflict between the "conceived" and "lived" spaces in the initial stages of the planning activity, in relation to Ankara city. The lands in the periphery of the city were considered as the green belt and recreational areas for the urbanites of the city in the context of strategical planning. In fact, these lands hosted large-scale agricultural activity as the main supplier of food for state institutions and tobacco for Tekel in the 1940s and 1950s. The first encounters of the rural-to urban migrants took place within the boundaries of the city in a context where "urban agriculture" brought paradoxical consequences for the modern urban plan in Ankara.

The reason behind the arrival of first migrant group from Kırşehir was to work as agricultural laborers in Üreğil villagers' farms and vineyards. This seems to constitute an important challenge to the conventional acceptance of the modernization approach that delineates agriculture and urbanization as conflicting activities. Interestingly, the relation between the rural-to-urban migrants and the native villagers in the periphery of the modernist city supplied the first context of the formation of the locality. The initial nature of the social and trust relations between native villagers of Üreğil and certain migrant groups had even defined power relations in the locality between different *hemşehri* communities in the coming decades. In the research setting, the main reason behind Kırşehir *hemşehri* community's appropriating certain critical spaces in the locality was their desire to benefit from building close relationships with the villagers. They were able take near-strategical decisions mainly in terms of choosing which lands to buy or appropriate, taking into account the natural risks of the time. They purchased the well-suited vineyards and farms of the Üreğil villagers at the best possible price after considering the natural risk.

As can be gathered from the narratives of the migrants, the development of locality as a *gecekondu* region had two important spatial causes. First, the presence of the railway connecting these peripheral lands with city center, and second, the presence of the Samsun Highway connecting the central Anatolian cities and towns with Ankara. In fact, the railway and highway were constructed as the outcomes of the modernist interventions of both *RPP* and *DP* in the city space respectively. However, these strategical interventions had led to the unintended growth of *gecekondu* settlements in Mamak. These two incidents of strategical intervention contributed to serious increases in land prices. Thus, agricultural lands were put on the market to be sold as *gecekondu* lands to the arriving migrants. The first amnesty law in 1948 triggered the *gecekondu* construction in the locality and Mamak in general. As was identified in individual narratives, a dense migratory trend to the locality had emerged by the 1950s.

The illegal process of *gecekondu* house construction witnessed the initial display of certain tactics by the migrants vis-à-vis the realm of strategy. The reports of the construction of houses reflect a game like relation with unwritten rules between the migrants and state officials, whose duty it was to apply demolition orders. In the narratives, bribing the officials in charge was the most often encountered tactic to protect the houses from demolition. Forming close personal relations with the officials, respecting the conditions of the unspoken rules of the game-like relation provide ways to protect the house from total demolition. As gathered from the descriptions of the demolitions, sometimes, in order to give the impression that orders had been carried out; officers would damage a part of the house in such as a way as not to make the entire house uninhabitable.

Certain tactics were employed by the migrants in the course of the construction of *gecekondu* houses; building *gecekondus* rapidly, in a single night, can be counted as the most important. Officials gave demolition priority to the houses under construction or those that had only been recently built. Another tactic was to construct the house initially, with the appearance of a different structure, by attributing different functions to it. As an example, most of the *gecekondu* houses under the "strict" and "selective" supervision of the time were presented as "coal cellars" to the officials, as a masking tactic. Within the context of this game-like relation, the officials overlooked houses constructed under this masking tactic.

The rapid pace of construction necessitated close community solidarity. Early migrants who established close *hemşehri* community in the region mentioned their advantageous position with regard to the community help both in material [providing for *hemşehri* resources which were limited at the time, for example, permitting them to use their wells, providing construction material and food] and non-material [mostly labor]. In the narratives of migrants it is also possible to understand how they defined *hemşehri* in this early settlement period. Unlike the conventional considerations in the literature, *hemşehri* refers to a very intimate group limited to relatives or at most people of the same village, as opposed to the migrants of the same province.

Settlement decisions also included some tactical instances especially for the early settler migrants who had the opportunity to take the decisions in the then underdeveloped space. The natural factors, the river being the most important one, represented important concerns in the decisions of the migrants about the most optimal location to settle. The initial settlers took their decisions considering mainly natural spatial indicators. For instance, they took into account the possible dangers (river-flooding had often taken place in the 1950s after heavy rains) and resources that the river possessed for construction (water and sand). As distinct from the lands near the river, the upper sides of the hills seem to be considered as unsuitable for settlement by the migrants in topographical terms.

The second important tactic for the migrants' survival in the "insecure" and "unknown" space of the locality was to settle close to their *hemşehri* community. Migrants from Kırşehir, who, due to their relatively longer-term experience in the locality, relative wealth and close social relation with Üreğil villagers, were able to buy large plots of land from the villagers to subsequently sell to *hemşehri*s.

Thirdly, migrants who were able to make predictions about the future appropriated large plots of lands that included state owned lands in combination with the lands they bought from the Üreğil villagers. By "colonizing the lands" of mixed ownership on which they built their *gecekondus*, they gained the right to state owned lands over time with amnesty laws and legalization decisions of the strategical realm. Particularly since the emergence of multiparty politics in the 1950s, populist policies of political agents in the strategical realm, including amnesty laws and legalization decision have combined, albeit hesitantly, with the degraded representation of these settlements and expressions of the urgent need to modernize them.

The near-strategical settlement decisions of the wealthy migrants to purchase land in places far from river, due to the risk of flooding, led to unforeseen consequences and missed opportunities for the migrants in the 1960s and 1970s. This was due to remarkable value increases in the lands immediately adjacent to either side of the river due to the construction of a market place there. Since the river provided the natural resources such as sand and water that were desperately needed for construction of *gecekondus* at the time, the first small workshops and a number of sand factories were built near the river in the 1950s.

This triggered the construction of the market place there. The wealthiest shop owners in the region who own the existing few apartment houses in the market area now, had originally been the poorest ones and the ones lacking *hemşehri* bonds in the initial settlement years, and who had no other choice but to settle near the river due to low prices of these lands. "X family" who has been identified as the most influential family in the locality and who had, to a large extent, determined the course of social spatialization in the market was one of these poor migrant families. *Kırşehirli* migrants, on the other hand, often mentioned their regrets regarding their choices of not settling close to the river, and some of them blamed the forerunner *hemşehris* for this "wrong" decision.

The following decade is specified as the most remarkable second moment in the history of the locality. In the 1970s, the social spatialization in the neighborhood had taken a radically political turn in congruence with the macro political and economic conditions of the country. Struggles between the rightist and leftist militant groups mainly defined the course of social spatialization on the horizontal level. The narratives of the respondents about this period include emotional depictions of the fights between militant groups, strong feeling of insecurity and sufferings especially of youths and male members of the family. These sufferings were described as a consequence of the continuous interrogations of militant groups' members. The *Sünni* and *Alevi* respondents alike mentioned their disapprovals of the struggles between militant groups that took place during this period. Yet, this period signifies the moment where they located themselves and their *hemşehri* community within the polarized identities of *Alevism* and *Sünnism,* which had become associated mostly with leftist and rightist ideology respectively. In the narratives, the sectarian identities of *Alevism* and *Sünnism* emerged as politically overloaded identities of this period.

The near-strategical acts of powerful *Sünni* shop owners had excluded mainly latecomer *Alevi* migrants from Çorum, Sivas, and later Tunceli from taking lands and shops in the most critical component of public space in previous decades. This affected the course of social spatialization in the late 1960 and 1970s. Unintentionally, the near-strategic attempts of the founders to control the social spatialization in the market place and general discomfort of *Sünni* shop owners with respect to *Alevis* could not prevent the place to turn into a stronghold of leftist militants in the late 1970s. This unintended consequence seems to have one important socio-spatial reason; exclusion of *Alevi* groups from the valuable central lands of the market place, and their concentration on lands close to Samsun Highway and the settlement of *Alevi* migrants of Tunceli on the upper parts of the hills nearby Samsun Highway instigated the close encounters between these groups and residents of Üreğil neighborhood who were mostly *Sünni* supporters of ultra nationalist militant groups on the other side of the highway. Therefore the train station, adjacent to highway which is located exactly in between these two neighborhoods, was described as the most crucial point in public space where open fights and even incidents of killings between leftists and rightists took place with reference to this period.

The spontaneity embedded in the social spatialization had been felt by the migrants to the fullest extent during this period due to the limited capacity and willingness of state security forces to intervene in the tactical realm. The relation between strategical and tactical realm on the vertical level seems to have been at its weakest point during those years. There was almost no reference to security forces and their intervention in the incidents and fights, in relation to this period. Since Boğaziçi was mainly known for its *Sünni* migrants from Kırşehir and Kırıkkale, the locality was not subject to the same total repression and interference by state security as had been the case in some neighborhoods totally populated by *Alevi* migrants. Due to the limited interventions of state forces, the shop owners who had the power to take effective near-strategical decisions in defining the course of social spatialization in the previous decades pointed out their greater sufferings because of the repression of leftist militants when compared to other shop owners. They had to use such tactics as closing their shops early in the afternoon, not staying in the shop without a companion or family members, and avoiding social gatherings that might be evaluated by militant groups as political gatherings. The pressure of militant groups, either by attacking shops or by collecting protection money, contributed to serious job losses during the late 1970s. Besides the insecurity, this is the most frequent complaint of shop owners related to the period. The shop owners' tactics for survival seem to be the main reason behind the job losses since this damaged the natural flow of socio-spatial relations in the market place. The restricted mobility of the residents because of the security tactics constituted another reason behind the job losses. They defined these years as their most difficult period in the region.

Since the interaction between the strategical and tactical realm during the late 1970s was weaker than all other periods in the locality, and the interaction between groups via the social spatialization of the neighborhood involved a serious dose of spontaneity, residents seemed to have found protection mainly within their *hemşehri* communities. In this respect, the prior settlement decisions in communities proved to be a successful security mechanism, particularly for the more homogenously and densely populated communities like Kırşehir and Kırıkkale. On the other hand, the *Alevi* and *Sünni* communities of heterogeneously represented provinces in sectarian terms, like Yozgat, Çorum and Sivas, mention their sufferings in relation to political incidents and tense relations with neighbors of the opposite sect because of their wide dispersal.

Mostly, the male migrants mention their sufferings as a result of continuous interrogations from both leftist and rightist militants on their way to work. Almost all of the male respondents identified one incident that they had experienced or that they had witnessed. Most of these incidents were narrated as having taken place at the train station. During these interrogations, one's hometown identity and, in relation to that, political affiliation seem to have constituted the core of the investigation. This is related to the fact that one's village of origin was often identified with one's sectarian affiliation. Within this context, depending on who was doing the interrogation some residents, as a tactic of survival, lied about their hometown identities.

The narratives of the 1970s, despite the highly chaotic nature of social spatialization, emphasize the importance of being neighbors. Being a good neighbor, as an important value embedded in *gecekondu* life because of socio-spatial factors, seems to have transcended the highly political nature of social spatialization at the time. Some *Sünni* residents in particular, narrate incidents where their *Alevi* neighbors, despite their political position, behaved protectively towards them. The good attitudes of neighbors from different sects seem to be related to both parties', in their own words, unwillingness "to behave shamefully" towards each other. The respondents, neighbors to a family of the opposite sect, gave almost no reference to any open struggle between them during these years. Rather, they, at times mentioned the fights of rightists and leftists that happened to take place outside the social and spatial context of these neighboring relations, even though one of the neighbors may have had militant involvement. At the extreme, in some few cases the relations between *Alevi* and *Sünni* neighbors may have lost its intensity or totally broken down due to the stereotypical prejudices and opinions, mainly on the part of *Sünni* neighbors. In cases where one or two families resided in a region totally dominated by the residents of the opposing sect, there emerged certain problems. I encountered a case during the research where an extended *Alevi* family felt the need to exchange their houses with a *Sünni* family residing in an *Alevi* dominated region because of the pressures of the *Sünni* community.

Sünni respondents mainly portrayed their *Alevi* neighbors as treating them well and as being protective during the period. This seems to be related not only to the good intentions of *Alevi* neighbors, but also for material reasons. *Alevis* had to -develop certain tactics of survival vis-à-vis the informing activities of their *Sünni* neighbors. *Alevi* respondents did not refer to any open conflicts between neighbors, however, after developing a certain trust relation; a number of them admitted incidents in which their *Sünni* neighbors reported them to the police in the late 1970s. Some *Sünni* women openly mentioned that they had informed on some of their *Alevi* neighbors to security forces for their illegal activities such as writing leftist slogans on the walls. Mostly, women referred to the nature of relations with neighbors in those years since they have always been the main carriers of the relationship with neighbors.

The third period, as specified by the narratives of the migrants, seems to have had peculiar qualities differentiating it from previous periods. This period in the 1980s is the one that explicates best the development of reciprocity between the strategical and the tactical realms.

Immediately after the military intervention, the main course of social spatialization was mainly governed through the repressive strategies of the military regime in order to suppress political activities in the locality. Three main strategies can be identified in these "depolitization" strategies. The first was to control the political activities in the locality by utilizing a secret information network, encouraging some migrants to report on other inhabitants in the neighborhood. Secondly, on the macro level, the neoliberal economic policies brought certain consequences that designated the economic concerns of the migrants as the major determinant of social spatialization rather than as political fights between militant groups. Legalization of *gecekondu* lands and houses, within the framework of the neoliberal policies of post military regime when connected with the reforms of decentralization in the local government system, not only brought a general increase in real estate values in the locality but also led to a more reciprocal relationship between migrants and local politicians. The new municipal system made local politics more responsive to the demands of *gecekondulu*. The increasing power of the migrants in influencing the strategical acts of developing plans or providing certain services to the locality seems to be the defining quality of this period.

Thirdly, as part of the strategic act of suppressing political activities in general, the military regime initiated, and succeeding civilian governments continued to implement the privileging of a *Sünni* interpretation of Islam as an antidote to the ideologies of Marxism and Kurdish separatism. In congruence with this strategy, the practice and effect of *Sünni* Islam become widespread in everyday life, especially in low-income settlements. Particularly after the mid 1990s, low-income settlements became the main supporters of the religiously oriented conservative political tradition. In parallel to these developments, the re-politicization of *Alevism* mainly took place with reference to the cultural and religious content of their identity with regard to the strategic policies of making the *Sünni* interpretation of Islam as hegemonic.

All these strategies led to a transformation in the course of social spatialization in the locality, from the "space of politics" to a "politics of space". The legalization of *gecekondu* neighborhoods, with the neoliberal policies of the 1980s, the increasing responsiveness of local politics to the demands of *gecekondulu*, the favoritist policies depending on the sectarian affiliation of the mayors in power, and the increasing religious claims of *Alevis* and *Sünnis* over the social space of the locality, further embedded politics in the social spatialization when compared to other periods.

Stories about local politics in conjunction with the macro policies of the time include a certain stress on the beginnings of a feeling of favoritism in the narratives of the migrants. In the aftermath of the highly political atmosphere of the late 1970s this period is remembered by the migrants with reference to the struggle to get legal right to their lands and houses with the best possible conditions vis-à-vis one's neighbors, and "other" groups. In relation to this, they carry an awareness of the possibility of partial treatment on the part of local strategical agents. The incidents of favoritism dominate the narratives of respondents in relation to the period, beginning in the mid 1980s and lasting until recent years. Favoritism in politics seems to have emerged out of the fact that the reciprocity and responsiveness of the relations between the strategical and the tactical realm on the vertical level intensified with the increasing power of migrants to define the fate of local politics. Therefore, the open struggles between leftists and rightists in the social space of the late 1970s transformed into a more implicit competition between *Alevis* and *Sünnis* or different *hemşehri* groups embedded more in the production of public space.

The social democrat municipal tradition in the 1980s that was followed by the Islamist conservative municipal tradition by the mid 1990s was narrated as favoring *Alevi* and *Sünni* groups respectively in the locality. Particularly, *Sünni* shop owners gave important references to their perceptions of favoritism during the social democrat mayoralty. The municipal police officers' intense and frequent supervision of the conditions of trading in the market place was the most often emphasized example of the biased attitude of social democrat mayors to the shop owners in the *Sünni* dominated market place. Powerful shop owners, through using their connections to critical posts in the strategical realm, also showed strong resistance against these policies.

The strategical realm, dominated by the neoliberal political agenda mentioned above, left many loopholes and uncontrollable spaces where tactical acts could be employed. The legalization process, through the agency of private construction offices under oath, brought about such uncontrollable processes. As an example, those who had no title deeds gained the right to the state owned lands on which they had built *gecekondu* houses. The decision depended on their verbal declaration about the size of the land they had appropriated previously.

Legalization, as a general strategy, brought unintended consequences, which in turn challenged its main objective, that is, the modernization of *gecekondu* districts. The legalization process and the integration of neoliberal consuming spaces in the market restored shop owners' power in economic and political terms because these developments contributed to increases in the value of real estate prices in the market place. Interestingly, this restoration of power on the part of shop owners contributed to the spontaneity of neighborhood space in terms of posing an alternative spatial development to the mainstream tendency of apartmentalization in low-income settlements. Due to the spatial necessity of initiating apartment construction at the edges of the main street in the market place, the shop owners gained a degree of control in determining the course of the apartmentalization process in the locality. However, shop owners benefited greatly from the legalization process because it protected the traditional structure of small *gecekondu* shops. Thus, shop owners were not so willingly to change the spatial structure of the market through apartmentalization because the rent revenues from these traditional *gecekondu* shops had increased with the rise in value of real estates. Therefore "slumlordism" for many shop owners was more beneficial than the gains they were supposed to get from the constructors in return for the apartments constructed to replace their shops.

The consumption patterns also changed with the opening of big supermarkets in the market place. The migrants often mentioned their preference for shopping for food from these supermarkets because of the attractive prices and the possibility of using credit cards. However, they also state that they did not want to offend their *bakkals* or other shop owners whom they had known for years because these shop owners still provided traditional and informal shopping opportunities of veresiye (shopping on credit) in cases of economic need. It is also necessary to note that *Alevi* and *Sünni* identities or *hemşehri* relations played a relatively minor role in determining the shopping choices of the migrants and the trust relations with the shop owners providing the opportunity of *veresiye*. Economic concerns, knowing each other for a lengthy peirod, and spatial proximity mainly defined the parameters of shopping concerns.

The third strategy of the state's imposition of a *Sünni* interpretation of Islam as an antidote to leftist politics contributed to the redefinition of *Alevi* and *Sünni* identities, mainly with reference to their religious content. The narratives in relation to this period cover such religiously-loaded definitions of *Alevism* and *Sünnism* rather than referring to the political connotations of these sectarian identities. Parallel with macro policies favoring *Sünni* Islam over *Alevism, Sünni* residents displayed an attitude of ignorance containing certain prejudices and superficial knowledge of the *Alevi* belief system. These negative value judgments about the *Alevi* belief system seemed to exist in hybrid forms alongside an appreciation of the good character or intentions of *Alevi* friends or neighbors who were well known. In general, the recognition and prejudices embedded in the distinction made between

Alevi and *Sünni* lifestyles dominated the expressions of the respondents with reference to recent social relations between different sectarian groups. In cases where trust between *Alevi* and *Sünni* neighbors had been greatly damaged in the late 1970s or through the informing activities in the early 1980s the relationship between neighbors was, at times, totally broken.

In general, it is possible to observe the increasing effect of religious concerns along with economic concerns in defining the course of social spatialization in the neighborhood. In this respect, the production of religious spaces has constituted an important part of social spatialization. There is no reference in religious terms to an open competition between sectarian groups over public space in the narratives of the migrants. However, there is a certain dislike, displayed in the narratives as an intimidating awareness or neglect about each other's religious spaces. While *Sünni* respondents seem to have been indifferent to the *Alevi* belief system and the *Cemevi* as the space of *Alevism*, they expected *Alevis* to play active roles in the production and recognition of *Sünni* Islamic space. Informal religious gatherings among *Sünni* women became widespread, despite the pressures of the security forces, mainly during the years of social democrat dominance in local politics but have lessened recently. These gatherings have replaced the traditional gatherings of women known as "gün". Therefore, mainly during this period, the possibility of *Alevi* and *Sünni* women coming together as neighbors became nearly impossible and the women of the two sects further isolated their relations with the repolitizication of sectarian identities with reference to their religious contents.

Unlike the nature of spontaneity that was mainly a consequence of the minimal intervention or non-intervention of the strategical realm in the 1970s, the spontaneity of social spatialization in the last two decades has been a consequence of a growing local agency that is able to take near-strategical acts and because of the increasing reciprocity between the strategical and tactical realms on the vertical level. This reciprocity, however, has functioned mainly to the advantage of those powerful groups with the capacity to take near-strategical acts rather than the migrants that ought to behave tactically within the context of the strategical realm on the vertical level and the near-strategical realm on the horizontal level.

The first of the main empirical conclusions that can be drawn from the research data is that the ethnic identities through sectarian and hometown affiliations constitute the main means by which migrants have developed certain tactics in dealing with the strategical acts on the vertical level and other tactical acts on the horizontal level. In this respect, *Alevism, Sünnism* and *hemşehrilik* have played vital roles in the development of migrants' tactics in conjunction with the political and economic context of the strategical realm, which have attributed these ethnic identities with changing values and definitions overtime. Secondly, the growth of the locality as a *gecekondu* region can not only be considered as an unintended consequence of urban development but also must be considered as a by-product of the sum of modernist strategical acts that have captured diverse and paradoxical acts and decisions within themselves. Thirdly, empirical data give important examples from the reciprocal and interdependent relation between the tactical and strategical realm. In other words, tactical decisions and practices employ the structural means or loopholes of the strategical realm rather than having an autonomous existence. An alternative perspective is that the strategical realm, to define and redefine its modernist project, always needs its "other". However, different agents of the strategical realm represent and define this "other" differently depending on the macro political and economic context of the country and socio-spatial transformations in peripheral settlements. Depending on the political and economic context, the agents who may have different agendas in the strategical realm may represent and deal with the tactical realm by using different means. Fourthly, the research data support the criticisms that were posed regarding the homogeneous and static definition of the strategical and tactical realm in chapter one. The actors in the tactical realm, depending on their relations with the agents in the strategical realm, can attain an active and determining agency in the social spatialization of the locality. This agency, on the tactical level, transcends the limits of tactical acts as defined by De Certeau. Therefore, they are defined as near-strategical acts throughout the empirical chapters, since they have the capacity to take decisions based on an assesment of the future and the production and employment of their space. This capacity not only affects the course of social spatialization in the locality but also leads to certain spatial and temporal moments of resistance against the strategical realm on the part of powerful migrants. As exemplified by the empirical findings, these acts of resistance emerged from the interventions of the strategical realm in the locality in terms of the application of legalization decisions, construction plans, provision of certain services or the legal supervision of political, commercial and real estate activity. On the macro level, these near-strategical acts also have transforming and determining capacities on the agency of the strategical realm, as can be best exemplified by the "post-planning" practices and policies. Fifthly, in the same way as the operation of the strategy-tactic on the vertical level, the interdependent relations between the near-strategical acts of the more powerful groups or actors and the tactical acts of the less powerful groups on the horizontal level may bring unintended consequences via spatial contingencies. These unintended consequences may, over time, alter power relations in the locality depending on the general social, political and economic context.

The empirical findings also support the premise that the main competition between the groups has taken place mainly with reference to the appropriation of and competition over space. This seems to be related to the general spontaneity embedded in the formation of *gecekondu* settlements. The identity definitions of *hemşehri* and sectarian groups have a dynamic nature that has been altered and contested depending on their position vis-à-vis each other in the course of social spatialization.

After making these general empirical and theoretical conclusions, I would like to mention the possible areas in which studies similar to this book may contribute. The belief in the need to increase the number of such studies constitutes the main inspiration. First of all, in today's cities, it is not possible to develop inclusive, participatory, democratic and successful planning and urban policies without local knowledge of social spatialization, particularly in relation to low-income settlements, including spatially, a serious dose of spontaneity. It seems highly important to understand the processes and patterns through which these localities have been produced through the means of a dialectical interplay between the strategical and tactical realms. This understanding may hopefully develop a new approach to urban politics that transcends the limits and impediments of conventional urban planning and governance understanding. It becomes even more crucial to develop a more inclusive and democratic understanding of local politics because of the operation of neoliberal political strategies which trigger the spontaneous nature of social spatialization in low-income settlements. This, in turn, dramatically increases the social and economic inequalities between more powerful migrant groups and late comer, "ethnically other" less powerful groups.

Secondly, space-sensitive methodological and conceptual approach used in this study may hopefully constitute an inspiration for other studies on low-income settlements. A spatial approach, or in Lefebvrian sense "spatiology", provides rich conceptual tools to understand everyday life or history in low-income settlements. On the theoretical level, the "social" and "spatial" cannot be thought of as apart regarding the theoretical and empirical discussions in this study. On the empirical level, "to have a space of one's own" constitutes the basis of everyday life because of the spontaneous nature of these spaces and the close encounters between different identity groups in *gecekondu* settlements. Therefore, spatial analysis is vital in understanding the social dynamics in the neighborhood.

It is also important to note that the close encounters between different *hemşehri*, ethnic and sectarian groups have not always resulted in conflict-ridden relations, as has been represented and exaggerated in public discourse for the last three decades. As the cases of neighbors from the opposite sects show, despite a continuous explicit or implicit competition over space that has changed its intensity and nature depending on the context, living close to each other for years and taking part in social spatialization in the region together, has brought migrants to an understanding and learning process about each other's life worlds. The findings of the research also give clues to the uses of spatial dialogue and the dangers of spatial fragmentation of different identity groups if one considers the socio-spatially created nature of social identity.

APPENDIX A

Questions Guiding the Organization of Semi-Structured Interviews, Informal Conversations and Observations with Regard to Neighborhood Inhabitants

Basic Demographic Concerns:

1 Name, Surname
2 Age
3 Education history
4 What is your occupation?
5 What are the jobs that you have done from the first years of your presence in the city until now?
6 What is the year of migration from the village/town to the city?
7 Where did you migrate from as a family?
8 What were the reasons behind your migration?
9 Where did you settle first in Ankara?
10 What were the different places that had been resided in Ankara in years?
11 When did you first arrive to this neighborhood?
12 Have you ever changed place/house in the neighborhood? For what reasons?
13 What were the factors behind your decision to settle in this particular neighborhood?
14 Do you own the house that you are living in?

In case they own the house:

15 How did you construct the house?
16 Did you get any help from anybody during the construction? If yes, what kind of help did you get?
17 Did you hire workers during the construction?
18 Did you get any financial help in the course of the construction? If you did, from whom did you get financial help?
19 What kind of difficulties did you encounter in the course of construction of your *gecekondu*?
20 What kind of reactions did you get from your neighbors or other migrants living in the close environment during your settlement?
21 From whom did you buy the land on which you constructed your *gecekondu*?
22 If you built your house on state owned land, who did play the mediating role in the appropriation of the land?

23 Do you have legal title deeds to the land?
24 Do you have construction permit for the house?

In case they own the house, but they are not the first owners of the house:

25 From whom did you buy the house?
26 Under what conditions did you make the payment?
27 Did you get any financial help from somebody?
28 Who was the first owner of the house? What was their reason of selling the house?

In case they do not own the house:

29 Do you reside in the house as tenants?
30 If not, under what conditions do you reside in the house?
31 Who is the original owner of the house?

General Questions Regarding the Early Settlement Period:

32 What kind of difficulties did you experience related to living in gecekondu in the initial settlement period?
33 Did you take any help from your neighbors for your immediate needs like water, electricity, tools and labor power in the initial settlement period?
34 Has your house ever been demolished during or after the construction?
35 If yes, what was your attitude in dealing with demolition?
36 When did the infrastructural services like water, sewer system and electricity become available in the region and for your house?
37 What did you do in order to demand or get these services?
38 Have you ever done any changes in the house? If yes, what kind of changes did you make in the house?
39 Do you own a house in other parts of Mamak or Ankara? If yes, how did you buy the house?

Questions Regarding Public Space:

40 How did you choose the place for settling down when you first came to the district?
41 In that period, with whom were you in relation in your near environment?
42 How was your relation with these people?

43 Where were they from?
44 How was the physical appearance of the district in that period?
45 Did you experience a disaster in the district in this period or later (like flood, landslide)?
46 If you did, how were you affected by this event or how were other people affected?
47 Was there any shop/store for buying food, clothes etc. when you first come to the district in the close environment?
48 If there were, which ones did you prefer?
49 What were your reasons for preferring these shops?
50 What kinds of shops were there in the district in the early settlement period?
51 If there were not such shops, where did you do shopping?
52 In case you were shopping from the shops in your near environment, were you paying cash or on credit?
53 If you were paying on credit, did you have any affinity with the owners of these shops? (in terms of hemşehrilik, neighborhood or being *Alevi/Sünni*)
54 How frequently and for what reasons were you going to the city center in the early settlement years?
55 What kind of transportation means were you using?
56 What is your opinion about the transportation system of the time? (What are their perceptions about the advancement of transportation over time?)
57 Which was the most difficult period for you in financial terms?
58 In these financially hard times, did you receive financial aid from anybody?
59 In the neighborhood, which was the hardest period regarding your social relationships?
60 With whom and what types of problems did you experience?
61 What kind of hardships did you experience in the neighborhood between 1975 and 1980?
62 Were you influenced by the political struggles of the late 1970s? How?
63 What is your opinion about the municipal services?
64 What can you demand from the municipality for your neighborhood?
65 What do you do in your spare times?
66 Where do you do your shopping for clothing needs?
67 If you do your shopping from the neighborhood, which shops do you prefer? For what reasons?
68 Where do you do your shopping for food?
69 If you do your shopping from the neighborhood, which shops do they prefer? For what reasons?

70 Where do you buy your other needs?
71 Do you go to mosque/*cemevi*? How frequently?
72 Which mosque/*cemevi* do you prefer? What are the reasons behind your preferences?
73 Where are the families living in your close environment from?
74 How are you relations with your neighbors? (the history of neighboring relations)
75 Do you have your village association in the neighborhood?
76 If yes, how was your village association established?
77 If yes, what kinds of activities are taken place in these associations?
78 If no, where is your village association in Ankara?
79 Do you participate in the activities of village association?
80 How frequently do you go to the city center and for what reasons?
81 Which hospital do you prefer to go when you are ill? What are the reasons of your preference?
82 Which schools did you attend for your primary and secondary education?
83 Was there any school in the neighborhood back then?
84 How were the schools constructed in the neighborhood?
85 How were the mosques constructed in the neighborhood?
86 How was the *cemevi* constructed in the neighborhood?

APPENDIX B

Questions Guiding the Organization of Semi-Structured Interviews, Informal Conversations and Observations with regard to Shop Owners

1 Name, Surname
2 Age
3 Education history
4 What are the jobs that you have done from the first years of your presence in the city until now?
5 What is the year of migration from the village/town to the city?
6 Where did you migrate from as a family?
7 What are the reasons behind your migration?
8 Where did you settle as a family first in Ankara? What were the different places that had been resided in Ankara in years?
9 When did you first arrive to this neighborhood?
10 Have you ever changed place/house in the neighborhood? For what reasons?
11 What were the factors behind your decision to open shop in this particular neighborhood?
12 When did you first open this shop?
13 Have you had another shop in the market place previous to this shop?
14 Did you construct the *gecekondu* shop by yourself?
15 If yes, how did you acquire the land on which the gecekondu shop was constructed?
16 How did you construct the shop?
17 During the construction, from whom did you get help in material or non-material terms?
18 If you did not construct the shop by yourself, how did you buy the shop?
19 How did you make the payments of the shop?
20 What kind of reactions did you get from other shop owners during the construction of the shop?
21 From whom did you buy the land on which you constructed your shop? If you built your shop on state owned land, who did play the mediator role in the appropriation of the land?
22 Do you have legal title deed to the land?
23 Do you have construction permit for the shop?
24 Which was the most difficult period for you in financial terms?
25 In these financially hard times, did you receive financial aid from anybody?

26 In the neighborhood, which was the hardest period regarding your social relationships?
27 With whom and what types of problems did you experience?
28 What kind of hardships has been faced in the neighborhood between 1975 and 1980?
29 Were you influenced by the political struggles of the late 1970s? How?
30 What is your opinion about the recently opened supermarkets in the market place?
31 How is your relationship with the municipal officers and zabıtas?
32 What do you think about the policies of local government?

The incidents that are taken into consideration during the participatory observations in the shop and informal talks with the shop owner:

33 What are the things sold in the shop?
34 How is the shop space used in daily life?
35 What are the conditions of shopping?
36 Does the shop owner sell the goods on credit?
37 If yes, how does the shop owner select the customers who are creditable?
38 Is the shop space used for functions other than trading like, social gatherings, association meetings etc.?
39 Is it possible to group the customers of the shop under one homogenous category with regard to the sectarian, hometown or ethnic identity?
40 Does the shop owner have customers from other neighborhoods?
41 If yes, how does the shop owner define the reason behind customers' preference of this shop in particular?
42 Is there any person or institution from which the shop owner gets material help or credit?
43 Is there any person or institution to which the shop owner gives credit or donations regularly?
44 What are the political inclinations of the shop owner?
45 Does the shop owner have a membership to any political party or organization?
46 Do the political inclinations of the shop owner have any impact on his selection of trusted customers?
47 Has the shop owner ever given reference to the religious issues in his/her narratives?
48 Has the shop owner ever given reference to the identities of *Alevism* and *Sünnism* in his/her narratives?
49 How does the shop owner define the nature of his/her relations with the neighboring shop owners?

50 Does the shop owner define any group or individual with whom s/he has an uneasy relationship?

51 Which aspects of the history of market place is emphasized more by the shop owner?

FSC
www.fsc.org
MIX
Papier | Fördert
gute Waldnutzung
FSC® C083411

Kolibri 360 GmbH
Ferdinand-Jühlke-Straße 7
99095 Erfurt, Deutschland
produktsicherheit@kolibri360.de

Druck:
CPI Druckdienstleistungen GmbH
im Auftrag der
Kolibri 360 GmbH
Ferdinand-Jühlke-Str. 7
99095 Erfurt